Exotic Foods

A Kitchen and Garden Guide

Marian Van Atta

Pineapple Press, Inc
Sarasota, Florida

Inquiries should be addressed to:

Pineapple Press, Inc.
P.O. Box 3889
Sarasota, Florida 34230

www.pineapplepress.com.

Library of Congress Cataloging in Publication Data

Van Atta, Marian, 1924–
     Exotic foods : a kitchen and garden guide / [Marian Van Atta].—1st ed.
          p.   cm.
     Includes bibliographical references and index.
     ISBN 1-56164-215-0
     1. Tropical fruit. 2. Tropical crops. 3. Cookery (Tropical fruit) I. Title.

SB359 .V35 2001
634'.6—dc21
00-061140   ус
                         0 = 44728146

First Edition
10 9 8 7 6 5 4 3 2

Design by *osprey*design
Printed in the United States of America

Contents

Acknowledgments

Many people helped to make this book possible. First the readers of my weekly "Living off the Land" column asked for more information. We then started the "Living off the Land Subtropic Newsletter." Readers who contributed articles, recipes, fruit and vegetables to experiment with and plants to grow are all part of this book. I am especially grateful to Kaye Cude, Bob and Opal Smith, Elizabeth Robinson, Lori Smith, Georgiana Kjerulff, Melvin Manthey, Sylvester Rose, Patrick D. Smith, Bill Bixby, Frieda Caplan, J.R. Brooks and Son, and Dr. Martin Price.

Betty Mackey grouped plants in logical order from our first 74 issues. Martha Draheim and Peggy Marion put the information on the Macintosh computer, and Anne Van Atta checked my second version. Dr. Franklin Martin helped with botanical names and technical advice.

Special thanks to Brigadier General D. E. Thomas of San Antonio, Texas, Gill Whitton of Tampa, and Tom MacCubbin and Robert Vincent Sims of Orlando, for promotional help with their garden radio shows.

We thank our subscribers, some of whom have been with us since the first issue in 1975. Without their continued support this book would not have been written.

Pineapple Press is making this book a reality—and we will no longer have to dig out copies of our first 74 issues. We thank them!

Florida Space Coast Writers' Conference made it possible for us to meet Pineapple Press. We thank Ed Kirschner for keeping this organization alive since 1984.

We thank the staff at Kwik Kopy in Melbourne for printing our newsletter and the Melbourne Post Office for sending it out all over the world.

I am especially proud of our contributors, who wrote from their personal experiences. They come from Florida, California, Arizona, Texas, Ohio, Haiti, Ecuador, and Germany. They helped me share ways for you to raise and use exotic foods.

Introduction

Welcome to the taste of sunshine! If you are a gardener who likes to cook (or eat!), or a cook who likes to garden, then this book is for you.

Here is a guide to the vast array of exotic fruits and other delicious edibles from the sunny tropics and subtropics. This book lists and illustrates more than 70 useful plants and gives growing instructions, sources, recipes, and fascinating facts about each one. Mango, pineapple, carambola, roselle, guava, Key lime, orange, blueberry, pomegranate, kiwi, banana, jujube, sapodilla, lychee, macadamia, olive, peanut, fig, papaya, cactus pear—these and many more fruits, nuts, and plants are included.

This book also tells how to find and use some of the easier to prepare wild edibles. For flavor, there is nothing like fresh-picked produce, wild or domesticated. It is an offering from the land that can easily be yours.

Though some of the plants described in this book may be new to you, it is not difficult to learn to know, grow, and use them. Like more familiar plants, the exotics share simple basic needs: fresh air, sunshine, soil with nutrients, warm weather, and water. Information on each plant type's size, soil preference, pest control, best varieties, and special needs lets you grow a lychee tree as easily as an oak, if your climate is right.

This book gives many workable ideas to gardeners living in the subtropics. In the United States this includes zone 9 (as designated by the U.S. Department of Agriculture), where minimum temperatures are 20 to 30 degrees, and zone 10, with 30 to 40 degrees, and the newly created zone 11, where temperatures do not drop below 40 degrees. The new USDA Plant Hardiness Map of the United States on page 2 shows areas where it frosts occasionally but rarely. The techniques in this book also work in corresponding regions around the world. With a possible global warming trend, and new tissue culture technology, some of these subtropical plants may be cultivated in other areas in the future. Except for the largest of the trees, the plants may also be grown in greenhouses, enclosed atriums, and sun-rooms.

In addition to providing exotic flavors, growing food in subtropical climates brings fresh food to your table year-round. This helps reduce the need for preserving. My husband, Jack, and I can easily feast on our homegrown produce and the fish he catches from nearby waters any time of the year.

Growing your own food assures you of purity and good quality in your family's diet. Even the most familiar foods, such as strawberries and oranges, taste better when fresh picked. And, because many food plants are attractive in the landscape, growing them enhances the value of your property. If more people grew fruit and vegetables, the food supply of the world would be increased. Insect control would be more manageable because plants would be spread out and less susceptible to disease. It is much easier to keep the pests off six citrus trees in your yard than off 1,000 trees growing in a commercial orange grove.

Step into your garden - what do you see? Are there exciting plants to look at, or some producing fruit? In my own garden, one of the finest sights is the tall carambola tree with golden star fruit. As space becomes more and more valuable and yards and gardens become smaller, we must become more selective about what plants to grow. Whenever possible, plants should serve more than one purpose. Many of the plants featured in this book offer landscape interest as well as flowers and fruit to bring indoors.

Many years ago I started to experiment with subtropical foods I grew at home or found in the wild. For over 20 years, my newspaper column, "Living off the Land," has been sharing what I've learned with readers of several newspapers and magazines. My readers—who include horticulturists, farmers, members of rare fruit clubs, garden clubs, and amateur gardeners—share information with me, too! My overflowing mailbag prompted me to start a newsletter, "Living off the Land" in 1975. Now much of this information, gathered by me and by my correspondents and researchers, has been assembled here in one convenient reference. Throughout the book I've noted when a section was contributed by someone else.

In many ways, this book is a trip around the world. The tropics and subtropics circle the globe, encompassing many cultures and nationalities. Latin American, African, Oriental, and Mediterranean food plants can share space in the same backyard. In many cases, the same food types are cultivated, but they are cooked differently by various ethnic groups. You will find some of these fascinating and delicious recipes within these pages. Some words of caution: the first time you try a new exotic food or wild edible, eat only a small amount. There is a slight chance you may have an allergic reaction. Some people are very allergic to mangos. Also be sure wild edibles are identified correctly by a local expert before you eat them. There are many "right" ways of gardening. Plants are adaptable to climate and shade, within limits, and soil builders and other methods are available for enriching garden soil. The instructions and recipes included here are based on the firsthand experience and practical advice of amateur and professional gardeners from a variety of climate zones. While this book will not eliminate trial and error in the garden or kitchen, it does offer many useful suggestions and ideas and will help prevent time-consuming, expensive mistakes.

The first chapter describes the needs of plants that grow in the subtropics. It explains how to enrich soil naturally with compost and cover crops. Basic principles of freeze protection, mulching, pest control, plant containers, and irrigation are discussed.

Succeeding chapters describe individual categories of food plants in detail. Each plant, its flowers, and its fruit are pictured together and described. Included are many types of citrus; nut and condiment trees; favorite American fruit trees especially bred for the subtropical climate; exotic fruit trees like loquat and carambola; shrubs bearing edibles, such as Surinam cherry; fruiting canes like raspberry and youngberry; plants with edible roots, such as water chestnuts; food-bearing vines, from grape to kiwi; unusual fruit from seed; and subtropical herbs.

Growing instructions and suggestions for using each plant are included in each chapter. There is also a separate chapter of recipes for jelly, jam, frozen ices, tea, wine, special calamondin cake and bread, and more.

The bibliography and list of sources for plants, found at the end of the book, will make the information in this book even more useful.

Many gardeners who are new residents of subtropical areas become discouraged when they try to grow Northern plants with Northern methods in this sometimes harsh climate, and find that their efforts fail. Hot-climate gardening is different, to be sure. But gardening in the subtropics and indoors is easy and rewarding if you choose from among the special plants described in this book. It is my wish and hope that readers will take advantage of the sun that shines year-round to grow delightful exotic and flavorful products of the warmer regions of the earth.

Growing Plants from Warm Climates

1

Plants are much like people. They have to be fed, watered, and kept from freezing. If you can take care of these basic needs, you will grow all the exotic edibles you can use. You can landscape your home and yard with exotic fruit-producing trees, bushes, and vines. Make them part of your life plan!

For more than 20 years we have been doing this. Jack says, "If it doesn't have edible fruit, I don't want to plant it!" But I have talked him into including some of the beautiful gardenias, jasmine, and hibiscus we couldn't grow in the North. We must all take time to smell the flowers too.

Freeze Protection

When you start to grow many of the fruits, vegetables, and herbs mentioned in this book, you will find most will not survive unless the temperature stays above freezing.

Most exotics thrive best in zones 9, 10, and especially 11, or in special microclimates (see Climate Map below). Zone 10 is almost a freeze-free area. Zone 9 has only occasional freezes and a growing season of 9 to 10 months. Zone 11, mostly in Hawaii and Key West, rarely drops below 40 degrees. Even if you don't live in these zones, you can still create your own microclimates by using protected areas. We plant our very tender tropical plants in a space between our house and our garage, which faces the sunny southeast. Take a lesson from coffee growers who plant quick-growing "nurse" trees to provide wind and shade protection for baby plants. You can grow castor beans and horseradish trees to protect guava trees, passion fruit vines, and other exotic fruit.

Zone 11
Zone 10
Zone 9
Zone 8
Zone 7
Zone 6
Zone 5
Zone 4
Zone 3

Many folks have kept their fruit trees from freezing by protecting them from the cold. If you have enough warm winters or can protect your tree until it grows higher than your head, it will often survive colder weather later on its own.

You can protect your plants from freezing in many ways. Use your old quilts, blankets, and drapes (especially the insulated types) as plant covers when frost is predicted. With the accurate weather reports and pictures from the TV weather satellite, we can now tell quite accurately when to expect a killing freeze. Palm fronds piled teepee-style around tender young trees also keep Jack Frost at bay. But beware of using plastic covers that come in contact with green leaves. Cold plastic kills! Save the big bats of pink fiberglass insulation to wrap papaya trees during freezes. It works! Also, don't forget to remove your plant covers as soon as the sun comes out and the temperature warms up, since covered leaves are quickly damaged. Dr. Jerome Keuper, founder of the Florida Institute of Technology, in Melbourne, Florida, used electric blankets to keep some of his baby palms from dying. The school now boasts an interesting palm collection, the second largest in the world.

Another method of protecting plants from freezing involves using water to coat the plant with ice. It sounds crazy but it works if done right. Because the temperature of ice is 32 degrees and plant tissue does not freeze until below that temperature, ice actually protects plants from freeze damage.

To protect our garden during a freeze, we make use of our deep well which maintains a temperature of about 72 degrees year-round, allowing us to irrigate our mini citrus grove and keep it from freezing.

A few years ago we had two nights when temperatures fell to 19 degrees. So we turned our garden hose on our carambola tree, which stands about 20 feet high. I have pictures of this tree, which survived and still gives us over 1,000 fruit each year, with 2-foot-long icicles on its branches. If you use sprinklers, you must keep them on until the temperature warms and the ice melts, or the branches may break from the weight of the ice.

Some folks have built "freeze protection" into their irrigation systems: they direct a spray of water on the trunk of the tree that keeps it from freezing. Treetops may be frost-damaged but if the trunk is spared, especially below the bud on grafted citrus, the tree will "come back."

Soil Enrichment

Two of the main requirements for growing exotic food plants in warm places (or for organically grown plants anywhere) are a composting program and a worm box. Both produce materials that aid in plant growth.

We have used many methods of composting, including a rotating drum which you can buy for about $125, bins made from cement blocks, wire rings made from chicken wire, and a simple compost mixture or "gardener's pie." We make our compost by piling layers of organic material—grass clippings, wood chips, pine needles, oak leaves, palm fronds, garden trimmings, and like items—in a circle. We add fertilizers and manures to hasten decomposition. (In most areas, riding stables will let you have all the horse manure you can haul.) We also add cheap detergents to act as wetting agents, and we keep the pile moist until it becomes crumbly. Then we add the finished black compost to all planting areas before putting in new trees, bushes, vines, plants, and seed.

Worm castings have been used for centuries to help plants grow. Rich in nitrogen, they were plant food long before the first commercial fertilizer was marketed. You can use a wooden box, garbage can, or even a plastic bucket to raise worms. (Make sure there are drainage holes in the bottom to keep the worms alive.) The container can be any size or shape. Use treated wood that won't decay when wet. Make a loose-fitting cover to keep out light and help retain moisture. Worms will die if they get too dry!

Order a start of earthworms from a worm dealer (listed in most gardening magazines) or stop by your local bait shop and buy a package of fishing worms. Put the worms in the worm box and add enough soil to fill the box halfway. Keep the soil moist but not soaking wet. Dig a hole in the soil in the worm box and bury your kitchen scraps each day. Cover the scraps well so they don't tempt local wildlife to come over for lunch and make a terrible mess. Occasionally, add small amounts of wood ashes, sawdust, or manures.

Recently we've put used paper plates, cardboard, and even junk mail into our worm box. The cooperative little fertilizer makers are helping to lighten our loads to the local landfill. They are the ultimate recyclers!

Cover Crops

Another way to enrich the soil in warm areas is to plant cover crops during the hot summer months when our growing of vegetables is limited to special items that can survive the heat. When your spring planting of lettuce, tomatoes, cabbages, and so on gives out, put the plant debris in your compost pile and lightly plow or rototill your garden area. Plant any of the following: amaranth, buckwheat, cowpeas, crotalaria, millet, peanuts, sorghum, or soybeans. Then, in late August, turn the cover crop under with your rototiller. Or if you wish to use a cover crop in winter, plant and turn under ryegrass.

Weed Control

My biggest problem growing anything in a warm climate is keeping ahead of the weeds. In Florida, we usually do not have a killing freeze to discourage weeds, so we find that when we clear off new planting areas they quickly become covered by these fast growing plants. Make it a practice to cover any cleared, exposed soil with some kind of a mulch. It will keep the weeds from taking over your garden. Of course you can eat the edible weeds. (I wrote *Wild Edibles: Identification for Living off the Land* at the request of readers of my newspaper column, who wanted to try some of the tempting weed recipes we've presented over the years.) But you probably can't eat them fast enough. We can never eat all the Spanish needles (*Bidens pilosa*) that come up in our yard—though we do enjoy them cooked with onions, bacon, and cheese. (Now we use bacon bits made from soybeans.)

But keeping the weeds from sprouting where you don't want them to grow is much easier than pulling them out. To prevent weeds from taking hold, we have used strips of carpet, tar paper, plastic, heavy cardboard, and the new woven-plastic weed block. All can be covered with ground pine needles and leaves to make the area look more attractive.

Indoor and Container Cultivation

If you are really serious about growing exotic foods, eventually you will probably build a greenhouse. A friend of mine who wintered in Florida spent her summers in Yellow Springs, Ohio, where she grew bananas, papayas, chayotes, and passion fruit in a greenhouse.

For many months of the year, a greenhouse in Florida will be much like a sauna bath. You need proper vents and fans to keep it from getting too hot.

Some very fine tropicals are grown in protected patios and swimming-pool enclosures in warm areas. And in the North, a sun porch is often used. A woman in Minnesota grew many passion fruit varieties in her yard and porch during the summer. In the fall she moved the plants indoors to the heated sun porch, living areas, and even the basement of her house.

In New York City, Debbie Peterson started planting avocado seeds in pots. The seeds soon grew to 6-foot trees. She met others growing avocados this way and eventually sponsored a contest at a flower show that drew hundreds of entries. Thus encouraged, she founded the Pits Society. Now members grow and have fruited in greenhouses or heated porches many of the same exotic bananas, papayas, citrus, and other plants that are grown outdoors in Florida, California, and Hawaii!

Some folks elect to plant their exotics in containers. It is a good way to make sure the soil is free of the warm-weather nemesis that hinders plant growth: a microscopic worm—the nematode. Use sterilized soil. Containers can be clay pots, wooden cases, metal tubs, plastic buckets, halved whiskey barrels, and the new, very large plastic pots big enough to hold 10-foot trees. Of course, the containers must have adequate drainage, and you must water them more frequently than you would plants growing in the ground.

Watering

The most important ingredient in successfully growing heavily producing, healthy exotic fruit and vegetables in the subtropics is an adequate water supply.

After my husband designed our first irrigation system of ½ inch rigid white plastic pipe, we struggled less to produce food crops. He always says, "Florida is alternately a desert and a rain forest." So when it's not raining we need adequate moisture. See the irrigation system for home gardens shown below.

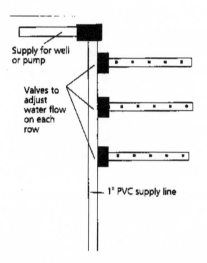

Supply for well or pump

Valves to adjust water flow on each row

1" PVC supply line

Semi-Drip Irrigation for Home Garden

Use ½" PVC pipe with 1/16" holes drilled 1' apart on center of pipe for row crops or drill a 1/16" hole at the base of each bush.

In 1957-58 we lived in Santa Ana, California, and we used city water. In California, Arizona, and even in Florida, where drought is now a problem, cisterns and water towers common to many rural areas and tropical islands can be used. A few years ago I had a letter from a Fort Myers reader asking for help in finding someone to rebuild the water tower of her 1900 house. Sadly, I couldn't help. But this skill should be revived. I was pleased to read that a house with a water tower was built recently in the beach area of Punta Gorda, Florida.

Adequate drainage is also important to prevent flooding in the event of heavy rains. Improve drainage by planting vegetable crops in rows with little ditches for runoff, using raised beds, and by filling very low areas with more soil before planting.

Landscaping with Exotic Edibles

Landscaping with food plants can be done easily. Many of the food-producers described here make fine focal points against the walls of your house. Instead of buying an expensive palm for the front of our house, we planted a loquat tree there. It is evergreen and produces tasty, nourishing fruit. We selected other edibles— including lemon grass, cactus pears, and aloes—to make a bed bordering the loquat tree on the street side of our house. Some cold-sensitive plantings like the Key lime, guava, and miracle fruit edge the southeast side of our house.

There is no end to the possibilities of landscaping with fruiting plants. Look around your neighborhood for good examples. Take courses from your local agricultural department or county extension office. Read and study books on tropical landscaping. Local newspapers and magazines often offer articles on the subject, along with drawings and pictures of attractive subtropical landscaping. Now "edible landscaping" is in vogue—and the way of the future. With renewed interest in conservation, it makes sense to plant edibles around your home and help save energy for all!

Citrus Fruit | 2

A great joy in living in the subtropics is being able to grow and use our own citrus fruit. Whether you live in the North or South, you can enjoy some of your own homegrown citrus. In cold climates you will thrill to the first bloom of your orange or grapefruit tree in your living room. You may not even mind hauling the plant outdoors for the summer (if you have put the growing pot on wheels).

In Florida, California, Texas, Arizona, warm spots around the Gulf of Mexico, Hawaii, and in warm countries around the world, you can grow some of your own citrus easily outside. Don't miss this special joy!

GRAPEFRUIT

Each morning from November to sometimes even the Fourth of July, we go out and get a fresh grapefruit from one of our 9 grapefruit trees. We choose one that has fallen to the ground and is in the perfect state of ripeness. Back in the kitchen I scrub the fruit, dry it, cut it in half, and put each piece in a bowl. Often I decorate it with a sprig of fresh mint or a sliced strawberry from my garden. Then Jack and I begin our day by scooping out and savoring each sweet, juicy segment. What a difference from "fresh" grapefruit from the supermarket when it has been shipped weeks ago, stored in bins until it is ready to mold, and at times picked too green in the first place. Only those who grow their own know exactly how delicious the truly tree-ripened grapefruit can be.

Grapefruit varieties you might like to plant are Duncan, a seedy white fruit that ripens from October to January; white-fleshed Marsh Seedless, which ripens from December to May; Red Seedless which ripens October to January; or Thompson Pink, which ripens October to January.

In July 1988, we attended the International Rare Fruit Growers meeting in California and tasted the exciting new Oro Blanco grapefruit. This hybrid of the Marsh and the low-acid Pummelo may soon be available to citrus growers in all areas.

A longtime former neighbor, Cap Fick, used to say, "Marian, never eat your fruit until it is ripe enough to fall on the ground." This, of course, does present some problems. Though some experts say never to mulch citrus, we bank ours with mounds of pine needles and leaves, and when the fruit falls it lands softly without splitting open. Of course we make sure the mulch doesn't touch the trunks, which could cause disease.

ORANGES and TANGERINES

When I was growing up in Milwaukee, my mother always bought California oranges. That was when Florida didn't have marketing regulations and sugar content wasn't monitored before oranges were shipped North.

Once we came to Florida and tasted truly tree-ripened oranges, we were pleasantly surprised. So we planted citrus trees in our first mini grove. I was surprised to discover that during the late spring and summer it was hard to find ripe oranges. We planted some summer-ripening orange trees, the late-blooming Valencias, to help extend the season.

Our favorite orange is the Navel. Jack's grandfather had a small grove in El Cajon, California, and his pride and joy was the huge, sweet Navel oranges he shipped in boxes to his grandchildren. We lived in Wisconsin then, and all the neighbors marveled at the fruit's perfection. I still remember the ambrosia, a Southern dessert my Alabama-born sister-in-law, Anne Van Atta, made from those Navel oranges.

But we have never tasted a freshly picked homegrown orange we didn't like. Any orange, including the old-time common Navel, Pineapple, Hamlin, or Parson Brown, makes wonderful fresh juice when the fruit is properly ripened.

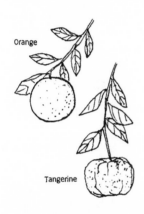

Orange

Tangerine

The best thing a Florida homeowner can do is plant orange varieties that ripen throughout the year. If you live in other citrus-growing areas, check with your local agricultural department for best-growing varieties and ripening dates. Ready first in October are the Satsumas, which are quite cold-hardy and grow in North Florida, around the Gulf, and in some Texas locations. Then come Hamlin, Parson Brown (October-December), Pineapple, and our Navels (December-March). Temple oranges and the easy-to-peel tangerines ripen January-March, and the Valencia variety ripens along with Lue Gim Gong oranges in March-July.

Other special orange-type fruits are the Mandarins (which include the Satsumas) and the tangerines. Our grandchildren love the zipper-skinned smaller fruit, which they could pick, peel, and eat as soon as they could walk!

Additional tangerine varieties and ripening times include Robinson, September-October; Lee, October-November; Osceola, November; and Dancy, December-February.

You won't want to miss having some tangelos, which are a cross between a grapefruit and tangerine. The Orlando variety ripens in November-January; the Mineolla ripens later, in January-March.

LEMONS and LIMES

While living in Wisconsin, never in my wildest dreams would I ever have imagined that someday I would have three kinds of lemons growing in my yard. But I do now: a regular lemon like you find in the grocery store, probably a Lisbon, planted by a former owner of our land; a Meyer, which looks a good deal like an orange and is quite cold-hardy; and a Ponderosa, which grows to the size of a basketball and produces enough juice for several pies. When we first moved to Florida, we labeled a Ponderosa lemon (as they did coconuts then) and mailed it to relatives in Milwaukee. It astounded the family and the mailman too!

Limes are another valuable subtropical crop. With today's low-fat, low-salt diets, it is nutritionally wise to flavor your foods with good-for-you limes.

One of our favorites is our Lakeland lime, a cross between a native Key lime and a kumquat. It survives our coldest weather and seems to produce its golf-ball-sized fruit almost year-round!

Then there's the Key lime (famous for pie-making), which comes in both a thorny and nonsticker variety. It is a lovely small tree, but it freezes easily at the first drop below 32 degrees. Our first Key lime did well and produced fruit much larger than you see in the grocery store.

However, we had to leave it at our first Florida homestead when we moved and later heard that it was killed by a freeze. Replanting in our new location, we put in a Key lime that bore only one fruit in seven years. We took it out and replaced it with another which was killed in the 1989 Christmas freeze.

Rough Lemon

Key Lime

But our Philippine Red lime, which is also called Rangpur lime, produces several crops each year. The tangy tartness from the small red-orange fruit is hard to beat.

CALAMONDINS and KUMQUATS

And then we have calamondins and kumquats. Both miniature orange-colored citrus relatives can add much excitement to your menu—and your landscaping, too.

Kumquats come in two types: one oblong and the other round. The oblong Nagami will make you wince if you pop one into your mouth before it is cooked and sweetened; the round Meiwa is less acid with sweeter rind that won't give you colic. One or two can be enjoyed right from the tree.

Calamondin

Both are usually small trees (though I did see some 20-foot calamondin trees at the Citrus Research Station in Lake Alfred, Florida). But the 3-foot calamondin laden with golden fruit growing in a pot in the front window of a Chinese grocery store in Lillooet, British Columbia, is a botanical treasure I'll always remember.

Did you know that many of the tiny "orange trees" sold in airport gift shops and citrus stands for tourists to take back home are actually calamondin trees? There are probably many of them growing in living rooms as beautifully as the one in Lillooet!

Other Citrus Fruits

There are still more citrus fruits you can grow. Ever hear of the Blood Red orange? We tried one, but the weather was warm then and this fruit seems to need some chilling to produce its purple juice!

Other special, very good-tasting citrus include the Honey Belle tangelos, the King and Queen oranges, the Poncans, and the Pummelos. The Poncans are a type of Mandarin with a puffy shape, and the Pummelos are bigger than grapefruit. This variety has firm flesh with little juice, so you peel it and eat the tasty segments. It comes with pink or white flesh.

And then there are the citrons. The Etrog is an oval-shaped fruit that has been used for centuries in Jewish religious ceremonies. Most amazing of all is the Buddha's Hand (or Fingered) citron, which is widely grown as an ornamental in China (though we have friends in Indian Harbour Beach who have a tree). Both are used to make traditional candied preserved peel widely used in fruit cakes. (There is also a citron melon used for candied fruit.)

Exciting new developments are taking place in the citrus world. Have you ever seen a citrus tree with several varieties grafted on it? Called the fruit cocktail tree, it is offered for sale in central Florida by a traveling citrus nursery.

In 1990 a hybrid orange, Ambersweet, was introduced. It took 26 years to develop this sweet orange crossed with a Mandarin and a grapefruit. It ripens from mid-October to December. Most importantly, it is hardy and may be able to survive freezes that have wiped out Florida groves in the past.

Lack of yard space should be no hindrance in growing citrus. Dwarf citrus are now available from California nurseries and Florida-grown dwarfs should be soon. The dwarfs include 6-8-foot Mandarin, Blood orange, and Tangelo trees; 3-6-foot calamondin and kumquat trees; 8-10-foot Navel and Valencia orange trees and grapefruit trees. Eureka, Lisbon, and Ponderosa lemon trees naturally grow only about 8-10 feet high, and you actually have to bend over to pick some of the fruit, which weighs branches down to the ground. Look for a straight trunk and clean union at the bud when choosing a citrus tree. We have had best luck buying citrus trees dug from the field. Ask your local nursery to order them for you. Instead of spreading granular fertilizers, which is heavy work, we have an easier way to keep our citrus trees well nourished. Every 8 weeks from mid-March until the first of November we use nutritional sprays to feed them—and they feed us in return.

Nut and Condiment Trees

3

Macadamia Nuts:
A Nutritional Treasure
Contributed by Myrtle Pell Edmon, Bellflower, California

Macadamia nuts, native to eastern Australia, are the most expensive nuts at the food market. They grow throughout the world from Africa to the southern United States. Hawaii currently produces the largest crop. California now has over 1,500 acres in commercial macadamia production. Some home growers in Florida are growing macadamias successfully. (We have enjoyed eating macadamia nuts grown on both coasts of Florida as well as those from Hawaii and California.)

An isolated mature tree will yield about 100 pounds of nuts each year. The nuts are a concentrated food low in moisture and high in fats, proteins, minerals, carbohydrates, and some vitamins. They are a particularly good source of calcium, phosphorous, iron and Vitamin B1. Raw nuts are palatable, but cooking enhances the flavor and makes them a nutritious snack or addition to any meal. First remove the outer husk. It's easiest to shell nuts after heating them in a 200-degree oven for about an hour. Apply the force of a hammer to the seam of the shell which you have set against a block on a rubber doormat-type pad. Some folks boil the nuts in their shells for 15 minutes. It makes them easier to crack without changing the flavor.

With 150 varieties to choose from, the one recommended for home planting is the hybrid known as Beaumont (660). It is a fast growing, upright tree producing enormous clusters of thinner–shelled, top-quality nuts. It reaches a height of 40 feet and a spread of 25 feet, but can be pruned to fit any landscape. Grafted trees can be expected to bear in 3 to 5 years while seedlings bear in 5 to 7 years.

Macadamias are resistant to root fungus and smog injury, immune to most diseases, and have few problems with insect pests. The tree has an exotic creamy bark with thick foliage. New leaves are reddish-bronze. In spring the tree is loaded with clusters of pink or white blossoms and in fall a crop of delectable nutritious nuts.

The easy-to-grow macadamia tree serves a dual purpose, for it is both ornamental and productive. The hollylike leaves are also used for decorations. Macadamia trees are sometimes available from local nurseries in the southern United States, but usually at prices over $20. If you want to save money you can grow your own from the seed (nut). Cut a small "x" on the bottom of each nut. Plant 1-inch deep in sand or vermiculite in full sun and keep moist. Seeds should germinate in 1 to 4 months. Cuttings from mature trees can also be rooted in sand, or you can air-layer from an established tree.

It is often impossible to find out the exact variety of macadamia nut that nurseries offer for sale. Gardeners will usually have to settle for an unknown variety. Nurseries should be educated to keep records of exact varieties. Macadamia nuts are botanically referred to as *Macadamia intergrifolia*.

Macadamia recipes

Macadamia Nut Brittle

1 cup chopped macadamia nuts
1–2 tablespoons butter
¼ teaspoon salt
2½ cups sugar

Prepare buttered plate or dish. Put nuts into saucepan and add butter. Cook until slightly brown. Add salt. In a very heavy skillet melt sugar slowly and stir to keep from burning. Pour over nuts on buttered plate. Spread quickly with spoon. Cool. Break into serving size pieces.

Macadamia Nut Pie

Contributed by Connie and Bill Schmitt, Melbourne, Florida

3 eggs, beaten
⅔ cup sugar
⅓ teaspoon salt
⅓ cup melted butter
1 cup light corn syrup
1½ cups chopped macadamia nuts

Mix all ingredients together and pour into pastry-lined pie pan. Bake at 375 degrees for 50 minutes.

Browned Macadamia Nuts

Saute macadamia nuts in a little butter or oil. Add a dash of sea salt and you have the high-priced nut you can buy in jars imported from Hawaii.

Hawaiian Fruit Salad

2 packages cream cheese
2 tablespoons sugar
½ teaspoon salt
Pinch cayenne pepper

Juice of 1 lemon
1 cup fresh grapes
¼ pound finely chopped macadamia nuts
2 tablespoons unflavored gelatin
2 cups water

Cream cheese until smooth. Add sugar, salt, cayenne, lemon juice, grapes, and chopped macadamia nuts. Soak gelatin in 1 cup water. Add remaining water. Heat to boiling and stir until gelatin dissolves. Set aside to cool. Pour dissolved gelatin over other mixed ingredients and stir thoroughly. Pour into a ring mold, cover, and set in refrigerator until firm. Serve on lettuce leaves with plain or vanilla yogurt.

Macadamia Nut Topping for Vegetables

Drain 2 cups cooked green beans, carrots, broccoli, cauliflower, or asparagus. Sprinkle with chopped parsley and a little lemon juice. Melt 1 tablespoon butter or margarine and add 2 tablespoons chopped macadamia nuts. Saute to a golden brown. Pour over vegetables and gently mix.

The Pecan:
A Fine Protein-Producer for the Subtropics

You can enjoy plenty of delicious nuts if you plant some pecan trees as part of your home landscaping. This native of North America is found growing wild in forested bottomlands of the Mississippi valley north to Illinois, west into Texas and Kansas, and at higher elevations in Mexico.

Most folks know of the large commercial planting of pecans in Georgia, Texas, and North Florida but don't realize pecans can do well in warmer areas. I personally know of a huge pecan, planted in downtown Melbourne in 1924, which yields steady crops each year. Laymond Hardy, in Miami, discovered two 65-foot pecan trees, planted in 1940, which produce bushels of nuts each year. Kay Cude reports a pecan tree that produces very large nuts growing in a very warm area on a riverbank in Fort Myers. Some pecans with very large nuts are also grown near Palm Springs, California.

There are more than 300 named pecan (*Carya illinoensis*) varieties, divided into three main groups:

Southern—has 200-day growing season and takes 40 plus inches of rain yearly.

Western—has 200-day growing season and takes 20 inches or less of rain.

Northern—growing season less than 165 days and is adapted to cold winters.

Proper planting of pecan trees is important to their success in your garden. Dig a large planting hole so roots can be spread out. The pecan has a long taproot so your hole will be quite deep. Plant pecan trees in a mix of peat moss and enriched soil, with an earth ring to hold in water around the young tree. Make sure new pecan trees get at least 10 gallons of water each week during their first year. They need a lot of water because of their very long taproots. Also, do not fertilize at planting time. Note, however, that Florida pecans often have a zinc deficiency. Apply 2 to 4 ounces of zinc oxide, spread evenly beneath the trees, each spring for the first two years.

Pecans lose their leaves each winter, so they are ideal trees for the west side of your house, where you want winter sun and summer shade. Plant more than one pecan tree for better crops. In central Florida, plant both Stuarts and Desirables in the same area.

Pecans contain 9.5 percent protein, 53 to 75 percent fat, and 10 to 15 percent carbohydrates, and are fairly high in phosphorus. They are a good source of vitamin A, thiamine, and riboflavin, and also contain lecithin.

Pecan recipes

Pecan-Carambola-Topped Baked Beans

> One 28-ounce can baked beans
> 3 slices bacon, diced, fried, and drained of grease
> 1 large carambola, thinly sliced (or use 1 large apple)
> ¼ cup brown sugar (or honey)
> ½ cup chopped pecans

Mix half of the pecans with beans and place in a 1½ qt. casserole. Top with carambola slices. Dot with butter. Sprinkle with brown sugar. Top with bacon pieces and rest of pecans. Bake at 350 degrees for 15 minutes.

Pecan Cheese Crackers

> 1 pound sharp cheddar cheese, grated
> 2 sticks butter or margarine
> 3 cups self-rising flour
> ½ teaspoon cayenne pepper
> 2 cups finely-chopped pecans

Cream butter and cheese together until well blended. Add flour and pepper and mix well. Add pecans. Shape into rolls, wrap in waxed paper and refrigerate 1 hour. Then cut ¼-inch thick and place on ungreased cookie sheet. Bake at 350 degrees for 15 to 20 minutes or until crackers hold shape. Do not brown. Makes 3 dozen crackers.

The Nogal:
A Subtropical Black Walnut
Contributed by Joy Hofmann, Lojo, Ecuador

With growing interest in tropical and subtropical tree crops, many previously overlooked species with economic values are being brought to our attention.

One is the nogal (*Juglans honorei*), a subtropical black walnut tree that is beautiful as well as useful. In landscaping it has proved hardy in center-lane highway dividers. It is grown in orchards where fruit is harvested for nutmeats. It is planted to check erosion and for fine wood crops. The leaves and bark are used for dyes and tonics.

In Ecuador, where nogal is native, it grows on cool slopes of the Andes mountains between 1,800 and 2,500 meters. It suffers no insect damage or disease. The tree grows from fresh seed. But 2 months after harvest, unplanted seed starts losing viability. Seedling trees begin to bear at 8 to 12 years. *J. neotropica* is the preferred rootstock for *J. regia* and other walnuts in warmer regions. The town of Ibarra, Ecuador, is known for little handmade wooden boxes made of nogal and filled with nutmeats and sweet candy paste. In another town almost the entire populace are wood carvers. Their hand-carved walnut wood statues and furniture are favorite collector's items.

(Cracking black walnuts is a problem. Women in Ecuador use rocks. Here, some folks spread unhusked nuts in driveways and run cars over them. Our Florida neighbor, who raises black walnuts on his Illinois farm, runs them through his corn husker, bags the nuts, dries them a short time, and then carefully cracks each with a hammer blow to the stem end. We have not been able to find reports of the tropical black walnut growing in the United States. But folks should try growing it. If you do, please let us know your results. M.V.)

Black Walnut recipes

James Beard suggested adding black walnuts to chicken and turkey salads, crepes, curries, and puddings. Black walnuts contain vitamins A and B, and iron. They are rich in unsaturated oils.

Black walnuts also make a superb ice cream. This recipe is from *Gather Ye Wild Things* by Susan Tyler Hitchcock, Harper & Row, 1980.

Black Walnut Ice Cream

Heat ½ cup maple syrup and 1 cup sugar to boil. Add about 1 cup nutmeats. Cool. Add 2 quarts milk, 2 beaten eggs and ¼ cup cream. Add a pinch of salt and freeze.

The Horseradish Tree:
Outstanding Nutritionally
Contributed by Lyle Ang, Haiti

The horseradish tree (*Moringa oleifera*), also called benzolive or drumstick tree, is a tropical tree that may reach a height of 30 feet in just a few years. It likes all soils and tolerates drought well.

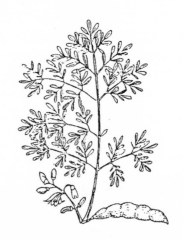

This tree is widely distributed throughout the tropics and is planted in southern California and Florida. Where frost is not a problem it provides excellent nutrition. Its seeds are easy to start, and the tree also grows well from cuttings.

The horseradish tree can be identified by its open crown of fernlike leaves with main leaf stalks up to 1½ feet long and thin elliptical leaflets about an inch long. Fragrant white flowers are about ¾-inch across in clusters on 6-inch stalks. Foot-long, brown seed capsules contain round seeds with 3 papery "wings." Older bark is whitish-gray and corky.

The leaves are the most abundant food part. Individual leaflets can easily be stripped from the tough stems by pulling them through the fingers. The more the trees are pruned the more abundant is the tender new growth. Only once did we pick a batch that was bitter, and they were older branches whose leaflets had started to turn yellow.

The leaves contain vitamin A (5900 International Units per 100g), vitamin B2 (9.4 mg/l00g), vitamin B3 (1.7 mg/l00g), calcium (264 mg/l00g) and iron (2.9 mg/l00g) with a well-balanced phosphorus content (81 mg/100g). To top it all off, the leaves are 38 percent protein.

The pods are also edible but must be picked when very young and easily snapped. Larger pods are very stringy, and although good eating, it's not worth picking strings out of your teeth.

Mature seeds can also be eaten but are very high in oil. This unique-quality edible vegetable oil does not easily become rancid and has superior lubricating qualities. The seeds are also used to purify water.

And about the name, horseradish tree? The roots are also edible and they taste like horseradish!

This tree grows wild in many warm areas especially in Hawaii and Haiti. Once identified, it is very easy to spot. The long pods hang straight down and are very persistent. Since the tree flowers throughout the year, there are always pods in the easy-to-see-through crown. It is much used as a living fencepost since it roots so easily from large branches.

Horseradish Leaf recipes

Lyle Ang, a forester in Haiti who furnished most of the information in this section, wrote: "Our household has been eating the leaves 3 times a week for the last several months. Our 3-year-old always cleans up the greens first! When simmered for 10 minutes or steamed in a pressure cooker for 5 minutes, they are very tender and sweet.

Young pods are supposed to have an asparagus flavor, but our family has not yet been fortunate enough to find tender pods. The abundant flowers make excellent bee pasture, and when cooked with the leaves, the flowers are also tasty. Herbs that go well with the leaves include tarragon, thyme, marjoram, and sweet basil. We usually eat the leaves as a 'spinach' and then add leftovers to soups and stews."

Benzolive Casserole
Contributed by Lyle Ang, Haiti
>   2 cups cooked leaves
>   1 cup milk
>   2 eggs
>   Onion, butter, salt, and pepper

Chop leaves in blender. In a mixing bowl beat eggs, then add finely chopped onions, butter, salt, and pepper to taste. Combine with the leaves and milk and pour into a buttered casserole dish. Top with bread crumbs and bake at 300 degrees until set (about 45 minutes). Sprinkle with grated cheese before serving if desired.

For further information on the horseradish tree, refer to the U.S. Department of Agriculture Handbook 249, "Common Trees of Puerto Rico and the Virgin Islands," 1964 by E.L. Little and F.H. Wadsworth. (Out of print but may be available at some libraries.)

The Versatile Olive

The beautiful olive tree is an evergreen with gray-green leaves. A *Olea europaea* is grown mainly in drier parts of zone 9 for its drupelike fruits and for landscaping. It is drought-tolerant and very easy to grow in the areas suited to it. The wood of the trees is tough, easily bent but not broken. Branches are numerous. The gnarled shape is part of the attractiveness of an older olive tree. In Spain and Greece, some trees are over a thousand years old. Olives have been cultivated since the beginning of civilization.

Most olive trees are self-pollinating. They bloom in April or May and the pollen is wind-borne. The olive will not fruit without mild winter chilling (ideally, 12-15 weeks of temperatures fluctuating between 35 and 60 degrees). Below 10 degrees, trees may not survive. Zone 9, with its long, hot growing season, is good for olives. The hot weather tends to promote oil production. Olive trees will not tolerate poorly drained soil, though they are not too particular about soil type and are tolerant of salts in the soil. As with other fruit trees, watering and fertilizing are good for fruit production. If humidity is too high, as it is in many Gulf Coast areas, the olive tree will grow but will not set fruit. Overly dry and windy weather will also decrease fruit set.

Girdling (cutting a thin strip into the cambium layer of the trunk) in December, January, or February increases the percentage of flowers and fruit yield. A nitrogen feeding in February stimulates more fruit the following year.

Varieties. In California, the popularity of the Manzanillo is spreading. The ripe olives of this variety are preferred by canners there. It ranks second to the Mission variety. Sevillano is third both in Spain and in California. Other varieties include Conasgueno, Real, and Morcal (late-ripening). In Greece leading varieties are Moretine, Vassiliki, and Amygdalolia. In Florida, olive varieties are most often unknown.

Characteristics. The leaves of the olive tree help it withstand heat and drought. They are opposite on the shoot and protected on top by a thick cuticle, and on the underside by matted hairs. Lanceolate leaves spread like rosettes and live for more than a year. Some varieties have leaves 3 inches long and half an inch wide; other varieties have shorter and narrower leaves. The underside is silvery, the upper side dark green.

Small whitish flowers appear in late spring on short panicles in the axils of a number of leaves along a shoot. Most are bisexual with stamens and pistil; some have stamens and a nonfunctioning pistil.

Propagation. From strong shoots, take cuttings 4 to 5 inches long, each with 2 leaves left at the top. They will root fairly well in clean sand if kept under high humidity. Rooting hormone helps. Cuttings should be from older branches with young tips removed. (The best time is autumn, after fruiting.) Another method is to soak cuttings for 24 hours in a solution of 13 ppm of IBA, bury in moist sawdust to callus for 30 days, then root as above.

Diseases. Olive trees are sometimes affected by olive knot, a bacterial disease. Cut out gall, then disinfect all wounds. The cut must be below the gall because the bacteria move downward.

Fruit. Olive fruits may be harvested long before they would drop from the tree naturally. They may be picked 6 to 8 months after blossoms appear, but they will hang on the tree much longer. They progress in color from green to straw to pink to red to black. They should be harvested green for pickling and black for pressing oil. (Contributed by Nick Acrivos, late founder of the Brevard Rare Fruit Council, Melbourne, Florida. Members have planted several olive trees in Gleason Park, in Indian Harbor Beach, Florida. Olive trees have produced large amounts of fruit in Brevard County, Florida.)

Cooking With Olives. Fresh olives are useful in every stage from immature green to ripe and black. But the same oil that makes them so valuable in the kitchen can be a slimy nuisance on the patio. Do not plant them where the staining fruit can fall on sidewalk or terrace. Olives and their oil give a smoky richness and saltiness to many dishes. Alone or mixed with any number of ingredients, they are a favorite in appetizers and canapes. Short of dessert, olives or olive oil can be found in every course or meal. Italian bread spread with olive oil and chopped fresh rosemary is great for breakfast! Then there are cream cheese and olive sandwiches for lunch, and all kinds of good things for dinner. Check any Italian, Greek, Spanish, or French cookbook.

Olive Oil. You can press your own delicious olives for oil. No need for waste; I've read that remaining pulp is fed to animals, and the liquid (after removal of oil) is an effective weed killer and insect repellent, particularly on cabbages.

Olive oil can be made in an olive or cider press. Dry olives (out of the sun) for a week, turning for evenness. Squeeze them in the press but do not crush the pits. Strain juice and allow to separate. Siphon off oil. Restrain the oil through cheesecloth every few weeks. The oil may be used throughout.

Brining Olives for Home Use. Nick Acrivos made these suggestions for brining olives: "To reduce bitterness, place green fruit in a solution of 2 ounces sodium hydroxide (lye) per gallon of water until the color change shows that the lye has penetrated to the pit. Then place in fresh water for 3 days or longer, changing water daily to remove lye. Soak 2 days in solution of 3 ounces sodium chloride (table salt, preferably iodine-free) per gallon of water. This will keep fruit for several weeks. For longer storage, change brine to 12 ounces salt per gallon of water for 3 days— add a little vinegar for flavor. Ripe olives develop color by being exposed to air between lye treatments." Caution: Lye is a caustic poison. Handle with care and remove from olives completely. Olives may also be brined in several changes of heavily salted water, which removes some bitterness.

Greek Chicken with Olives

Contributed by Betty Mackey, Longwood, Florida

> 8 floured, peppered chicken thighs
> 1 tablespoon olive oil
> 16 ounces stewed or fresh peeled sliced tomatoes
> ½ teaspoon marjoram
> 1 minced clove garlic
> ½ cup white wine or chicken broth
> 1 cup quartered mushrooms
> ½ cup sliced, pitted green or black olives

On high heat, brown 8 floured, peppered chicken thighs in skillet coated with 1 tablespoon olive oil. Turn heat low. Add the tomatoes, marjoram, garlic, and wine or broth. Cover and simmer half an hour. Add a cup of quartered mushrooms and ½ cup sliced, pitted green or black olives. Simmer 10 minutes and serve over rice. Garnish with minced parsley. Serves 4.

Pine:
A Remarkable Tree

How about a full-course meal served from a one-stop supermarket right in the middle of a forest? Everything—soup, nuts, spaghetti, bread, cooked vegetable, delicious tea, and even an after dinner mint—is included. Afterwards a very comfortable bed and some cool afternoon shade will be provided.

Our pioneers and tribes of native Americans learned to use many of the pine tree's bounteous gifts for their comfort and survival. Pines served us for centuries, providing shelter, fuel, furniture, food, and transportation. Our North American continent is endowed with rich forests of pine. The trees feed our woodland wildlife, serve in many ways, and even shed their foliage to carpet the forest floor. It's hard to imagine a world without pines. Pines come in many varieties. Depending on method of classification, up to 300 species of pines are found throughout the world. Our subtropical forests boast many varieties of the proud, bountiful pine, including the majestic longleaf, Cuban, slash, scrub, sand, Caribbean species, and pinon pines of the Southwest. Often we find pines growing next to palms.

Many pines have an inner bark which may be cooked or eaten raw. Needles can be steeped in boiling water to make a tea rich in vitamin C. Young pine cones can be ground up and used to flavor soups and stews. Pine nuts (seeds) hidden in the cones are delicious and nourishing. Our American gum chewing habit was adapted from early chewing of pine gum. Some Native Americans cut the inner bark into spaghettilike strips and cooked it with meat and vegetables. Some pounded the bark into cakes which were wrapped in moist leaves, baked slowly, and smoked for several days. Some tribes used pine sap to relieve colic and softened the bark in water to use as dressing for wounds. Pioneers ground the inner bark and mixed it with flour, harvested the nuts, and used the young needles for tea or just plain chewing. New England Shakers made candy from young pine shoots boiled in maple sugar.

When you take bark from pines or any trees, be sure to cut it off from small branches rather than from the trunk, to avoid injuring the tree. Pine tree products can be helpful to gardeners. To prevent cornworms from damaging corn, mix a tablespoon of a pine-based cleaner in a bucket of water. Either spray or pour a few drops on corn plants when they are 3 or 4 inches high. Use regularly until harvest. Pine scent keeps bugs away. Use pine needles as mulch in your garden. Put a light covering over newly planted rows of seeds. Watch the new plants sprout up. The pine covering helps keep the weeds down and is not too acid for most plants.

Cooks can use the male aments (flower clusters) of the pine, which appear in early spring. They look like small yellow cones and are clustered in the axil of the needles near the branch ends. Each sack is filled with pollen. Parboil first to get rid of excess pitch. Then put in a pan around roasting meat to finish cooking. (Don't try this if you have allergies to pollen.) If you make pine needle tea, add a few orange peels. This helps the flavor and adds vitamins, too.

Caution! One member of the pine family, the Yew (*Taxus torreya*), has red berries which are poisonous. It contains a powerful alkaloid heart depressant. Never eat Yew berries or seeds!

Favorite American Fruit Trees

4

Apples for Warm Climates

Contributed by Mary Ellen Smith, Melbourne, Florida

Only a Northerner could know what burns within me and how tears come to my eyes when I see apple blossoms and smell what can only be spring in our hearts, here in Florida." (From a letter I received from a reader in Miami. She wrote about the first apples from her Anna tree in Florida.)

When fruit as common as the apple is able to be grown in a warm area, it becomes an "exotic" fruit for those who grow and eat it. An old dream is coming true: apples are now growing in Florida and other warm areas! Three varieties—Anna, Ein Shemer, and Golden Dorsett—have proved viable as dooryard trees. Southern

Pride, Amy, Granny Smith, and others are being tested. The Golden Dorsett began as a freak. Irene Dorsett of Nassau, the Bahamas, bought a Golden Delicious apple in New York, took it home, ate it, and planted the seeds. One seed sprouted and grew into a tree that 6 years later bore fruit, in Nassau! William Whitman, Bal Harbour realtor and founder of the Rare Fruit Council International, discovered the tree and brought cuttings back to Miami. From these came the Golden Dorsett, flavorful and delicious and much like its ancestor.

The Anna and Ein Shemer varieties came from Israel. Anna is large, red, crisp, and tart. Ein Shemer is smaller, round, and sweeter. Experts recommend planting the Ein Shemer or Dorsett with the Anna to ensure cross-pollination.

But no one in Florida should jump into apples in a big way, at least not yet, cautions Dr. Robert Knight of the USDA Research Station in Miami. "There is much to be learned," he said.

What is known so far can be summarized as follows: select sturdy trees in cans. Plant two varieties blooming at the same time. Add peat moss to the bottom of the planting hole. Use well-composted manure as top dressing. Fertilize young trees every 2 months, bearing trees 4 times a year. Use 8-3-9 and once a year 6-6-6. Apply minor elements in a spray. Spray for aphids only if necessary. Be sure to prune after each crop and water liberally. Picking off all the old leaves every fall seems to help speed fruit production.

Apple trees take coddling, though cold won't hurt them. The apple is not yet a subtropical commercial crop, but once several problems are overcome, apples may become a dooryard staple in the subtropics. (Low-chill apples do much better in California and there are many more varieties to plant there including: Anna, Dorsett-Golden, Mutsu, Empire, Melrose, Fiji, Braeburn, Waltana, and Granny Smith. These will provide apples from May to February.)

Apple recipes

Mary Ellen Smith continues: "Early settlers in Florida, like my parents who moved to Melbourne in 1908, missed apples and made futile attempts to grow them. But I remember my mother's apple dumplings, made with store-bought apples, as one of our favorite desserts. Each child was served apple dumplings in a soup dish with hard sauce or cream." Here is her recipe:

Apple Dumplings

    6 tart apples, peeled and sliced
    1 teaspoon lemon juice
    About 1 cup sugar (less depending on tartness)
    ¾ cup water
    1 tablespoon butter

Combine in kettle and let the mixture set while fixing dumplings.

    2 cups flour
    3 tablespoons sugar
    1 egg
    3 teaspoons baking powder
    Milk (enough to make a stiff batter)

Sift flour, sugar, and baking powder into mixing bowl. Add egg, mix well and then add enough milk to make a stiff batter. Heat apple mixture on stove and bring to a boil, stirring often. Drop dumpling batter, a large spoonful at a time into boiling apple mixture. Cover with lid and cook for about 15 minutes. Serve with hard sauce, cream, milk, ice cream, or yogurt.

Applesauce

Wash, quarter, core, and peel 2 pounds apples. Bring ½ cup water to boil in a large heavy saucepan. Add apples and steam until soft, about 15 minutes. Run apples through colander or food processor.

At the Van Atta homestead we planted 1 Anna, 2 Ein Shemers and 1 Golden Dorsett tree. We often washed them off with "city water" from our garden hose and never had insect problems. However, the trees only lived for 10 years and then died from fire blight. But we haven't given up on apples. Last spring we planted a Southern variety of Granny Smith, the green apple from New Zealand. We decided there might be some connection between loquat fire blight and apple decline. It would be wise to plant your apples as far away from loquats as possible.

Please Pass the Peaches

For 9 years in a row, we had a bushel of peaches each growing season from our Florida Belle peach tree planted in 1979. We also had several dozen red-blushed fruit from our Florida Beauty planted in the same year.

Growing peaches by the subtropical homeowner has not always been easy. Horticulturist Arnold Pechstein of Sebastian, Florida, gives this explanation: "Until the Ceylon peach was brought from China and combined with nematode-resistant rootstock to make many new warm-weather peaches, growing peaches was a hit-or-miss situation." Now, thanks largely to the work of R.H. Sharpe and others at the University of Florida, we have a nice choice of peaches we can grow in warm places.

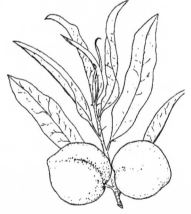

Peaches can be grown in a variety of soils. We think ours did very well on former piney-woods acid soil. We mixed compost, peat moss, and manures for the planting hole and kept the trees well mulched and watered during our terribly dry summers of 1981 and 1982. We used no poisonous sprays, but we did have some Caribbean fruit flies at the end of both crops. In the future, we hope to have a beneficial wasp-release program to avoid any fruit loss.

We were advised to prune our peach trees in January, but by then the tree was covered with small fruit and bright pink blooms. So we skipped that year and later pruned several weeks after picking the last fruit. We fed the trees in February, May, and August with composted chicken and horse manures and dried seaweed.

We were especially pleased with the large size of the white-fleshed Florida Belle. The Ceylon peach, which was introduced to Florida in 1880, has fruit the size of ping-pong balls and reddish flesh. We sampled fine Ceylon peaches grown by friends in Fort Myers.

Home growers should check the chilling requirements of peach varieties to fit each locality. Here in east central Florida, we get from 100 to 200 hours of under 45-degree weather each year, while south of Lake Okeechobee, there are only 50 to 100 hours of chill yearly.

Peach trees usually live from only 7 to 10 years and then need to be replaced.

Our California friends can easily grow many peach varieties with fewer problems than we have in Florida. In fact, California furnishes peaches to the whole country each summer!

Peach recipes

No fruit is better than the peach when picked fully ripened right from the tree and eaten immediately. If the birds will allow you this privilege, great! As the subtropical peaches produce their whole crop in a few weeks, you may have to preserve some. I simply peel, slice, and put in a small container, with or without a bit of honey, and freeze. Peach slices are delicious dried in any of the new dehydrators and make a good food for camping. Here are a few special peach recipes from Ruth Horomanski, Satellite Beach, Florida:

Peach Crisp
    4 cups peaches, peeled and sliced
    1 cup all-purpose flour
    1 cup sugar or honey
    1 teaspoon baking powder
    1 teaspoon allspice
    1 large egg

⅓ cup melted butter or oleo
½ cup chopped walnuts

Line a greased 8x8x2-inch baking dish with peach slices. Sift dry ingredients together and work in egg with pastry blender until mixture is like cornmeal. Sprinkle over peach slices. Drizzle butter and nuts on top. Bake at 350 for 45 minutes or until brown.

Low-Calorie Peach Smoothie
1 cup skim milk
1 cup sliced peaches
1 tablespoon lime juice
1 cup crushed ice

Chill all ingredients. Combine in blender and process a few seconds. Serves 2. Ruth uses mangoes for these recipes too.

Peach Pops
Peach slices can be frozen on a cookie sheet. When hard, remove and store slices in plastic freezer bags.

Pears for Warm Climates

Do you know pear trees live for 200 to 300 years? If you plant pear trees now, they may be feeding folks in the year 2200!

Several of our neighbors in Melbourne Village have grown pears successfully here in zone 9. Some of my college students have also brought in beautiful, tasty, homegrown pears from this area.

Experts claim most pears do better north of Orlando, but the Pineapple,

Hood, and Flordahome pear varieties are being tried farther south.

Pear trees can be grown on a variety of soils but do best in fertile, sandy loam with good drainage. In Florida, pear trees usually grow to about 20 feet high and about 25 feet wide. They do best in full sun and will not tolerate salty beach areas.

Field-dug pear trees are best planted while dormant in early winter, but potted pear trees can be transplanted almost any time. Make sure the planting hole is large enough to keep the root system from being crowded or broken. Leave a ridge around the edge of planting hole to form a water reservoir. Don't fertilize at planting time.

Use an organic 6-6-6 in January and again at the start of the rainy season in June. About 1 pound for each year of age of the tree is recommended until a maximum of 10 pounds is reached. Excessive fertilizer makes pears more susceptible to fire blight.

Irrigate trees during dry periods. Thoroughly wet all areas under the tree's canopy to a depth of several feet. Young trees may need 5 to 10 gallons, while large trees may need more than 500.

Prune yearly to remove dead wood, which helps open the tree and encourages branch spreading. Use mulches to keep area under trees weed-free. We always try to grow our fruit trees without poisonous sprays. Just washing fruit trees with "city water" helps keep them free from disease.

Most pear varieties are self-fruitful. Exceptions are Flordahome and Pineapple, which require cross-pollination. It is a good idea to include at least one Hood pear when you plant these varieties as they will all bloom at the same time and you will then have more fruit.

Pear trees are deciduous and lose their leaves each winter. They are cold-hardy and might be the answer for folks who want to replace a mango tree that freezes back too often with a good-producing fruit tree.

Pear trees are often available at local nurseries. A good mail-order source for warm-weather pear trees is Hastings. They carry Pineapple, Orient, Keiffer, and Ayers for zone 9, and Moonglow, Seckel, and Ayers for zone 8. See Appendix 3.

Watch for the big round "Chinese Pears" at your local supermarket. They are distributed by Frieda of California. Wonder what would happen if you planted the seed? These round "Pear-Apples" grow well in some warm areas.

Trees of the Shinseiki and Twentieth Century Pear-Apples are available at several California nurseries. See Appendix 4.

Pear recipes

Blanche and Ralph Boardman live in a modern-day log cabin in Palm Bay, Florida. They planted several "old timey pear trees" when they bought their land. Now they make chutney from their own fruit for very special Christmas gifts. Here's their recipe:

Blanche's Pear Chutney
>   7 pounds of pears
>   2 pounds seedless raisins
>   6 cups sugar
>   2 cups vinegar
>   Grated rind and juice of 3 large oranges
>   1 teaspoon each: cinnamon, cloves, and allspice
>   1 cup broken pecans

Chop pears using meat grinder or food processor. Add everything except pecans. Bring to a boil and simmer 2 hours. Add pecans and return to boil. Pour into hot sterilized jars. Makes 12 pints.

Pears Poached in Wine
>   4 firm pears
>   2 cups dry red wine
>   Juice and grated zest of 1 lemon
>   ¼ cup honey
>   1 teaspoon cinnamon
>   8-ounce package cream cheese
>   ½ cup chopped pecans

Halve pears lengthwise and core. Peel, leaving stems intact.

Combine wine, lemon, honey, and cinnamon in saucepan. Add pears. Simmer 20 minutes. Drain liquid (use in fruit drinks). Chill pears. Make filling by combining 8 ounces cream cheese, enough pear juice to soften, and ½ cup chopped pecans. Spoon into pear halves. Garnish with fresh mint.

Fresh Pears with Dips

Cut fresh pears into wedges. Serve with your favorite hot pepper dip or make your own by mixing ¼ cup pear chutney with 4 ounces cream cheese. Try combining 2 tablespoons thawed or mashed fresh strawberries with 4 ounces softened cream cheese for a pretty pink dip.

Dried Pears

Quarter washed pears. Slice thin for fast-drying pear chips. Cut in halves for slower-drying larger portions. Use dehydrator, solar or oven type.

Pear Leather

Cut pears in halves. Core and chop. Puree in blender. Add just enough water to make it run. Spread out sheets of plastic wrap, held down with masking tape, on dehydrator trays. Spread pear puree on the plastic about ¼-inch thick. Run dehydrator until leather dries. Takes about 2 days.

JAVA PLUM:
The Tree You Love to Hate

Unless you have actually tried to grow the Java plum (*Syzygium cuminii*)—sometimes called Jambolan—you cannot imagine how a tree can affect your life.

This member of the myrtle family grows faster than almost any other tropical fruit tree and will quickly grow higher than your house. The luxuriant evergreen foliage produces deep shade. The tree is drought-resistant, will grow in almost any type of soil, and will actually live in low places where it stands in water for months at a time. It is also disease- and insect-resistant.

The tree will soon take up most of your yard (our Java plum tree took a 60-foot square). It also puts out new trees from the roots and quickly sprouts more trees from fallen fruit. You soon have a thicket that you may not be able to control. Another big problem for folks like us who live where we sometimes get freezing tem-

peratures is that great portions of the Java plum will die from cold weather. Picking up the dead wood and constant pruning to keep the wood from falling on your house can be quite a chore! The 1-inch pear-shaped fruit is usually astringent when raw. It reminds me of a little purple eggplant, though it grows in grapelike ciusters. Never plant a Java plum next to your house or near a driveway, because great quantities of the fruit will cover the ground faster than you can pick it and will stain everything purple. However, Java plum trees make good windbreakers. After they are established, they need no aid in growing. The bark, seeds, and leaves are used medically in India. The bark is also used for tanning. The wood is so strong that it is used for making tools and carts. It is also used for fuel.

Java Plum recipes

Java Plum Juice
> 8 cups ripe plums
> 1¼ cup water

Wash fruit and remove thin green stems. Place in kettle with water and cook until fruit is soft. Pour through cloth jelly bag and drain off juice. I pour cold juice into freezer containers to keep in deep freeze. When ready to serve I add honey or sugar to taste. Other fruit juices, such as grapefruit, guava, apple, and mango, can be added.

Java Plum Gelatin

> 3 cups Java plum juice
> 2 envelopes plain gelatin
> ½ cup honey
> Juice of 1 small lime

Soften gelatin in about ½ cup cold water. Put 1 cup plum juice in saucepan. Heat and add softened gelatin. Add remaining juice and honey. Mix well. Pour into bowl and chill. Serve with plain yogurt.

Java plums make wonderful wine. This was the first tropical fruit we used for making wine. See recipe in Chapter 12.

Persimmon:
A Worthwhile Fruit

Nick Acrivos was a valued researcher and consultant. He contributed the following to our "Living off the Land" newsletter:

"We have found Japanese persimmons do not deserve the bad reputation given many persimmons. This is because some folks do not know when fruit of an astringent variety is ready to eat. A persimmon, when soft to touch, is ready! If it is a nonastringent variety, it may be eaten when a touch of color appears. Some of these are even eaten green. Our experience began with the purchase of a 2-year-old grafted Tanenashi persimmon. It grew slowly. The second year it had a large number of fruit; however, a large number aborted.

"Until then we watered all trees once a week. When we changed our system to twice a week the next year, the tree produced 35 large fruit.

"Since then we have increased our persimmon trees to 15, including the following varieties: Armond, Hanafuyu, Hachiya, Saijo, Hanagosho, Kungsunban, and Gionbo."

Nick was excited about budding persimmon trees. He first grew seedlings of the Kaki persimmon and the native Virginian, wild in Florida. But he preferred the Kaki because of the native's habit of sending out suckers.

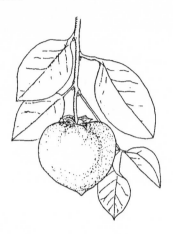

Seeds germinate readily in an equal-parts mixture of peat moss and vermiculite. Nick used a liquid fertilizer in conjunction with osmocote. Plants are watered every 3 days, as growth seems to cease if the plant becomes dry. Persimmons enjoy a moist environment, especially when in flower or fruiting. Nick budded persimmons in the spring, although budding can be successful at other times of the year.

The persimmon loses its leaves in early fall. Many trees have been disposed of during this period by owners thinking they were dead. But persimmons break out with new leaves in early spring. Persimmons grow throughout Florida and around the Gulf, and do exceptionally well in California, where they are grown commercially. We recommend that you plant a persimmon in a sunlit corner of your yard—and start enjoying the wonderful fruit.

Persimmon recipes

Persimmon Puree

Persimmon puree is used in many persimmon recipes. To make, wait until fruit is soft and almost jellylike. Use a spoon to scoop out flesh. Discard tough skin, seeds, and stem. Press through a food mill or whirl in blender. Fruit with tender skins can be cut in quarters, seeded, and whirled in blender.

To freeze persimmon puree, add 1 tablespoon lemon or lime juice to each cup of puree. Pack in freezing containers. Cover and freeze. (When in a hurry, wash persimmons, dry, and freeze each fruit in a plastic bag. Later, cut fruit when partially thawed.)

Persimmon-Grapefruit Appetizer
Alternate persimmon wedges with grapefruit sections in serving glasses. Chill and serve, garnished with fresh mint.

Persimmon Leather
Wash fully ripe persimmons. Cut in chunks and run through blender. Add honey to taste if not sweet enough. Pour puree in well-oiled baking dish only about ¼-inch thick. Bake at 300 degrees for 30 minutes. Turn off oven and keep persimmon in oven until dry. It should peel out of dish easily. Store between layers of wax paper in a covered container.

Persimmon Shake
Put 8 ounces low fat milk in blender. Cube and add 1 fresh persimmon. Whirl until fully mixed. Makes a wonderful thick shake that serves 2.

Elderberry:
A Wild Edible for All Seasons in the Subtropics
Contributed by Donald Ray Patterson, Titusville, Florida.

Many wild edibles are so diversified in their offerings to man that it is difficult to single out their best advantage. So it is with the colorful and profuse elderberry, which has showered us with gifts of food, medicine, tools, and handicrafts for many centuries.

This heavily fruited tree or shrub grows in abundance from Alaska to Newfoundland and throughout most of the United States, from California in the west to Alabama, Georgia, and Florida in the south. The elderberry prefers a rich, moist soil and is most frequently found along washes or streams from the flatlands to the lower mountain elevations.

Elderberry leaves grow 1 to 3 inches long in pointed, toothed leaflets. The stems are woody, containing a white pith. They dry and become hollow with age. Elderberries boast lush blossoms that grow in cream-colored sprays or cymes. In most locations, they flower and bear fruit between March and August, although the berries have been found year-round in Florida during mild winters.

As a food source, elderberries offer considerable variety. Juice, jam, jelly, and wines are made from the fruit while the flowers are used for tea and baked goods. As a note of caution, however, the red berries borne by some elderberry species are reported to be poisonous and should be avoided. However, black elderberries, *Sambucus nigra,* which we find in great quantities in Florida, are edible. They do not have a rank taste that requires drying the fruit before use, but can be picked and prepared in many ways. We think they should be cooked and not eaten raw.

Elderberry has been used in folk medicines for many years. Poultices of leaves treat sprains and bruises, while the inner bark is fried in fat and used as a skin ointment. We have used hot elderberry tea as a remedy for stomachache, sore throat, fever, and the common cold.

The value of elderberry doesn't end with food and medicine. It continues to serve us as an attractive ornamental, being used in many parts of the country as a flowering landscape shrub. For all its beauty and utility, the elderberry is a splendid bargain in nature's warehouse of wild edibles.

Elderberry recipes

Elderberry Pie
Contribution from Joyce Orcutt, Punta Gorda, Florida
> 3 cups elderberries, fresh or frozen. (To freeze, wash
> and drain. Pull from stems, pack in containers,
> and freeze.)

¾ cup sugar or ½ cup honey
1 tablespoon lemon juice
2 tablespoons flour

Place berry, flour, and sugar or honey mixture in 9-inch pastry-lined pie pan. Dot berries with 1 tablespoon butter. Moisten edge of crust with water. Put on top crust and crimp edges with fingers. Cut slits in top and brush top with milk. Place in preheated 400-degree oven. Bake 45 minutes or until browned.

Elderberries make great wine. See the recipe below and in Chapter 12.

Elderberries were once thought to be the homes of witches, and anyone who cut down the plant invited bad luck. Elderberry bloom and berry pickers were always sure to thank the witch of the elder!

Donald Ray Patterson, who wrote about elderberries for our newsletter, included his favorite recipes:

Elderblow Tea #1
Soak elderberry blossoms in cold water for 24 hours. Strain and serve hot or cold with lemon and honey to taste.

Elderblow Tea #2
Steep elderberry blossoms and mint leaves in boiling water for 15 minutes. Strain and drink hot or cold.

Elderblow Wine
Mix 1 quart elderberry blossoms with 3 gallons of water and 5 pounds of sugar. Add 1 package of dry yeast and allow to ferment for 9 days. Strain mixture and add 3 pounds of raisins. Age in a cask in a cool dark place, and then decant by siphoning, and bottle.

Elderberry Flower Waffles or Pancakes
For waffles, wash blossom dusters, shake dry, dip in waffle batter, then deep fry to a golden brown. For pancakes, strip florets from stem and mix with your favorite pancake batter.

Elderberry Chutney

Crush 2 pounds of cleaned elderberries and mix with 1 chopped onion, 2 cups vinegar, 1 teaspoon each of salt and ginger, 2 teaspoons honey, and a sprinkling of pepper and mixed spices. Bring to a boil and simmer, stirring frequently. Pour into jars and seal.

Elderblow Muffins

Mix together 1 cup biscuit mix, 1½ cups elderberry blossoms, 1 teaspoon shortening, 1 beaten egg, ½ cup milk, and ½ cup orange juice. Fill greased muffin cups ⅔ full and bake at 400 degrees for 20-25 minutes.

Elderberry Jelly

Crush 3 pounds elderberries in large saucepan. Heat gently until juice begins to flow. Then cover and simmer 15 to 20 minutes. Let cool. Using a jelly bag, squeeze out 3 cups juice. Put in saucepan with ¼ cup lemon juice. Mix in 1 box commercial pectin. Bring to boil while stirring. Add 4½ cups sugar and boil for 1 minute, stirring constantly. Remove from heat, skim off foam, and pour into hot sterilized jars, and seal.

The Marvelous Mulberry

Originating in Persia, the mulberry has spread all over the world. The mulberry (*Morus sp.*) is a member of the *Moraceae* family and is related to the fig, breadfruit, and rubber trees. Mulberry trees grow well in warm climates and produce large amounts of fruit. Because the fruit is very delicate, it is not suitable for shipping and is grown mostly for home use.

In Florida and Hawaii, the black mulberry is most commonly

grown. This picturesque, deciduous tree usually grows up to 30 feet (and sometimes taller) and has a short, rugged trunk branching to form a broad, irregular dome-shaped crown. Flowers appear as small catkins and form oblong fruit that is green at first, then pink, then finally a glistening purple-black.

Mulberries can be propagated from seeds, cuttings, and airlayers. In autumn, after the leaves fall, take 8-inch cuttings from strong new growth and place in wet sandy soil. Next spring, when roots have formed, put into pots. Then in the fall, plant in a permanent location where the trees will get some sun and good drainage. But remember to select a site well away from your house, driveway, or parking space. The ripe, juicy fruit stains easily. You can also start cuttings in spring before trees bud out.

Superior varieties are Hicks, Stubbs, and Townsend. Several other varieties include Russian mulberry (*Morus alba tarica*), a very hardy white mulberry; Mongolian mulberry (*Morus mongolica*), a native of Korea and China that is a shrublike tree growing to 25 feet with sweet, pale red fruit; and the Korean mulberry (*Morus acidosa*), which grows only to 10 feet and has dark red, sweet, juicy fruit.

The Wild Mulberries

A favorite food of the Indians, the native American red mulberry (*Morus rubra*) grows wild from New England to Florida and west to Ontario and south to Texas. The leaves are fine-toothed, somewhat sandpapery above and hairy below. The trees grow to 60 feet, and the fruit is red at first and then purple when ripe.

The Texas mulberry (*Morus microphylla*) grows from Texas and Arizona to Mexico. It is a shrub or small tree seldom growing larger than 12 feet high. It bears dark, pleasant-tasting fruit.

The white mulberry (*Morus alba*) was introduced in the United States by the British just before the Revolution in hopes of setting up a silkworm industry. White mulberry leaves are hairless, and the bark is yellow-brown. The white fruit, sometimes with a purple tinge, is sweet but rather insipid. It grows wild in many mild areas of the world.

David Fairchild wrote about white mulberries from Afghanistan. The fruit is almost seedless, and is dried and eaten by workmen who have nothing more except water while working a full day. The berries must have been very high in nutrition!

In Hawaii, a black mulberry was introduced for silk culture sometime before 1870. It became naturalized in various parts of the islands. Because of a lack of cross-pollination, it often appears in a seedless form there. It is an excellent, well-flavored fruit.

Several years ago we discovered several wild red mulberry trees growing in Florida. The fruit is tasty and can be used just as you use the cultivated mulberries.

Mulberry recipes

You can create your own pies, cakes, jelly, jam, and preserves from your mulberries. To avoid the tedious clipping of the tiny green mulberry stems, run mulberries through a blender and use the resulting puree for pies and sauces.

One-Crust Mulberry Pie

> 4 cups stemmed mulberries
> ½ cup sugar
> 2 tablespoons lemon or lime juice
> ¼ cup cornstarch
> ½ cup water

Add sugar to mulberries in a saucepan. Mix cornstarch with water and lemon juice. Add to berries and cook over medium heat until thick and juice becomes clear. Pour into baked pie shell. Cool and serve plain or with vanilla yogurt.

Easy Mulberry Wine

Pick and wash 6 cups mulberries. Cook with juice and peel of 1 lemon and 2 cups water. Pour through jelly bag. Heat and add 10 more cups water. Add 5 cups sugar. Stir until dissolved. Cool and add 1 package baker's or wine yeast Pour into a glass gallon jug and put on air lock. Ferment (usually takes about 2 weeks in Florida, but could be longer if weather is cool.) Remove air lock from jug and bottle if you wish or just put original jug lid back on. But be careful when you pour off the wine not to mix the sediment in from the bottom.

Exotic Fruit Trees

5

The Avocado:
An Aid to Good Health

With food costs continuing to go higher and even possible shortages foreseen for the future, anyone living off the land in the subtropics could provide some personal insurance against starvation by planting some avocado trees. When hurricane food shortages occur, a supply of homegrown avocados could come in handy.

Avocado fruit is high in food energy. It contains some protein and fat as well as calcium, phosphorus, sodium, potassium, vitamin A, thiamine, riboflavin, niacin, and ascorbic acid. Fourteen minerals including iron and copper are found in avocados. Avocados are best eaten fresh, but they can also be used in bread, cake, ice cream, pies, and main dishes.

Avocados come in three races: West Indian (large fruit with shiny, smooth, medium- to dark-green skin), Guatemalan (medium-sized fruit with green, purple, or black pebbly skin), and Mexican (small, smooth-skinned, purple or black fruit). West Indian avocados grow in hot humid tropics, while the Mexican and Guatemalan varieties grow in cooler highland areas and are more cold-hardy. Many varieties have been developed for commercial and home use. Besides the standard green, avocados come in many colors ranging from the black California-grown Hass to the mangolike pink blush of the Fairchild.

William Krome, longtime Homestead, Florida, avocado grower and researcher, recommends the following avocados for an extended harvesting season in areas with a South Florida-type climate: DuPuis, Bitte, Booth 7, and Kampong. Alternate choices include Pollock, Simmonds, Wilson, Peterson, Tower 2, Nordoy, and Nelson. For colder areas as far north as Valdosta, Georgia, plant Mexicola, Topa Topa, Duke, Young, Gainesville, and CRC 1411. Because pollination is sometimes a problem, it is best to plant several varieties that bloom during the same season.

When you plant an avocado seed it will probably take about 6 or 7 years to bear fruit. Budded trees, which have a bud from a fruit-producing tree implanted, usually bear fruit in 3 years. If you have only a small area to plant avocado trees, you could graft more than one variety on a single trunk, which could lengthen your harvesting season and help with cross-pollination.

Avocados grow well in full sun on well-drained soil but need adequate feeding. Avocados require more nitrogen than do most other fruits. One-half pound of fertilizer for each inch of trunk diameter every 3 months is recommended for young trees. The fertilizer should contain the minor elements. Animal manures are recommended. Scatter fertilizer around the trunk as far as limbs extend.

It's difficult to tell when avocados are ripe enough to pick because the fruit stays green. Some growers depend on feeling the fruit with their hands. It is best to know the approximate ripening date of your fruit. Do not allow your tree to become too tall or harvesting fruit will be a problem. Prune avocados right after harvesting. Our biggest problem has been the squirrels that chew on the avocados as soon as they begin to ripen. When we see the first "bushy tailed rats," as Jack calls them, eating our avocados, we try to harvest as many as we can. Some take a week or more to fully ripen indoors.

Here's a tip for planting an avocado pit: when the main stem is 6 inches high, cut it back to 3 inches. Otherwise your tree will be straight and spindly instead of developing offshoots and a uniform leaf structure.

Avocado Recipes

In case you have more avocados ripening than you can eat, you can freeze some for later use by following this recipe:

Frozen Avocado Puree
    4 ripe avocados, medium size
    ½ cup lime juice

Halve avocados and remove peel, or with a spoon scoop pulp from shell into a bowl. Sprinkle with lime juice. Mash or blend until smooth. Pack into glass jars or plastic freezing containers, leaving 1 inch air space. Seal and freeze. Remove from freezer, keep fruit in same container and allow to thaw in refrigerator about 24 hours. Keep tightly covered until used. (One of my readers puts a layer of mayonnaise over top of mashed avocado before freezing. It keeps the fruit from turning brown.)

Avocado Soufflé
    1 medium avocado
    2 tablespoons butter
    4 tablespoons flour
    ½ cup milk

½ teaspoon each: salt and pepper

3 eggs, separated

Peel and mash avocado. Melt butter, stir in flour, then milk, and make a thick white sauce. Cool. Stir in egg yolks and beat well. Beat egg whites until stiff and fold into sauce mixture. Fold in avocado. Add seasoning. Pour into a greased deep dish. Bake at 350 degrees until risen and golden brown (about 30 minutes).

Avocado Cake

1 cup pureed avocado

½ cup sugar

½ cup butter

2 eggs, well beaten

½ teaspoon each of cinnamon, nutmeg, allspice, and salt

1½ teaspoon baking soda

⅓ cup buttermilk

½ cup chopped dates

¼ cup raisins

½ cup chopped walnuts or pecans

1½ cups sifted white flour

Cream together sugar and butter. Add eggs and mix until light and fluffy. Add spices and baking soda and mix well. Add buttermilk, dates, raisins, and walnuts and mix well. Add flour and mix again to make a stiff batter. Spoon into a 9x9-inch greased baking pan. Bake at 325 about 1 hour or until toothpick comes out clean.

Our Favorite Guacamole

Mash 2 avocados well with a fork. Add 1 or 2 seeded and chopped green chiles. Mix in 2 tablespoons lime juice and about 1 teaspoon salt. (Omit salt if desired) Add chopped ripe tomatoes and finely chopped fresh green onion, crumbled crisp bacon, chopped roasted peanuts or pecans. Beat in a touch of olive oil, cumin, and finely chopped garlic and cilantro, or fresh parsley. Delicious used as a dip with corn chips or added to seafood, chicken or vegetable salads.

More Ways to Use Avocados

Add cubes or slices of avocado to scrambled eggs. Add avocado just before eggs are set.

Sprinkle diced avocado over top of tomato soup or add avocado puree to hot soup.

Marinate avocado slices in French dressing. Use as garnish for roasts, meatloaf, and fish.

Stuff celery stalks with mashed avocado seasoned with lime juice and garlic salt.

The Incredible Carambola

Averrhoa carambola is one of the most unusual and easily grown subtropical fruits. The fruit looks almost like wax and hangs heavily like decorations on a Christmas tree. Northern visitors are always awed!

The fruit is thin-skinned, yellow or cream-colored, with a distinctive 5 or 6-angle cylindrical shape about 3 to 5 inches long. Sometimes called starfruit, carambolas are very juicy, but the fruit from some varieties is sweeter than others. There are also some sour varieties which a few people like. When found in stores they are usually not labeled. You may get sweet or sour.

The blossoms, a light purple or pink, are often borne right on the main trunk as well as on the branches. Trees are moderate-sized, evergreen, and have delicate compound foliage.

A native of Indonesia, the carambola is very popular in Hawaii and Florida. It also grows well in Central America. California Rare Fruit Growers are trying to find the best way to grow it in their special microclimates. Varieties include Newcomb, Golden Star, Pei Sy Tao, Thayer, Arkin, Ti-Knight, and others. Carambola trees can be ordered from many local nurseries in Florida. You will probably be able to find them at local Rare Fruit Club tree sales too.

Propagation is from seed, air-layering, and grafting. A well nourished carambola tree will produce fruit almost year-round with large crops in early spring and summer.

Carambola grows well in sun or partial shade but will not tolerate salt. It prefers acid soil, but does well with heavy feeding of composted manure and grass clippings with the addition of minor elements. It must be kept well watered. We have experienced no insect problems with our carambola trees.

Young trees must be protected from cold. Established trees may be damaged at about 27 degrees, but after pruning the damaged limbs from our tree after the disastrous cold winter of 1977, our carambola recovered and produced one of its very best crops. It also survived the severe freeze in 1989.

Carambola juice contains about 10 percent natural sugar. It is a good source of vitamin C and contains some B vitamins and vitamin A as well as calcium, phosphorus, and iron. But because they contain oxalic acid, too many unripe or raw carambolas should not be eaten, especially by folks prone to gout! The green fruit has a high potassium oxalate content. The juice can be used to remove rust from fabric and will burnish brass.

We always wait until our carambolas are fully ripened and fall to the ground before harvesting for our personal use. We enjoy sharing a tree-ripened carambola sliced in star-shaped sections for lunch.

Carambola recipes

Carambolas can be made into jelly, jam, pickles, and preserves and can be baked in pie. We discovered that after freezes when the fruit has fallen to the ground and has turned a bronze color, it can still be used to make juice, jelly, and wine. In fact, carambolas make wonderful wine. One time we bottled it just right and it tasted like champagne.

See Chapter 12 for recipe.

Kathy's Carambola Bread

¾ cup pureed carambola pulp
1¾ cups flour (part wheat germ)
2 teaspoons baking powder
¼ teaspoon soda
⅔ cup sugar
⅓ cup shortening
2 eggs
½ cup coarsely chopped pecans

Sift flour, baking powder, and soda together. Combine sugar and shortening. Beat until creamy. Add eggs and beat well. Add flour mixture alternately with carambola. Add nuts and mix again. Bake in oiled loaf pan (8x4x3) at 350 degrees for about 1 hour. Loaf is done when a sharp knife inserted into center comes out clean.

Carob:
The "Chocolate" Candy Tree

Originating in the Holy Land, the carob (*Ceratonia siliqua*) grows in the same climates as the orange. The leatherlike edible pods can be made into flour. From it many delicious chocolate-type candies, pastries, cakes, and drinks can be made.

The carob is an evergreen tree growing up to 50 feet. It has pinnate leaves of 4 to 6 shining leaflets up to 4 inches long. It grows well on drained soil.

Carob has posed some interesting questions for me. Florida botanists I consulted insisted carob is dioecious, requiring both a male and female tree to reproduce. However, John Riley, co-founder of the California Rare Fruit Growers, writes in his book *Growing Rare Fruit in Northern California,* published by California Rare Fruit Growers, 1972, "Plant carob varieties which are self-pollinating." Later I discovered a mature carob tree blooming in my neighborhood. We found pods under the tree, but they were too dry to eat. As far as we can determine it is the only carob tree in this area.

Florida scientists say carob pods usually rot on the tree because of our wet weather. Perhaps with our increased dry cycles, carob may someday produce edible pods in Florida—if it has not already done so.

According to my research, the female or bisexual carob produces flat brown pods. They can be eaten "as is," but don't try to crack the seeds with your teeth! The stonelike seeds were the original "carats" used by goldsmiths. Now carob seeds are heated and made into tragacanth gum used for sizing cloth and added to inks and cosmetics.

Carob comes in numerous varieties, including Amele from the Adriatic, Tylliris from Cyprus, Santa Fe from California, and many more. Carob trees have been growing wild for centuries in the Middle East and are known there as St. John's Bread, Locust Bean, Honey Bread, Algarroba, and Caroubier.

The U.S. Patent Office distributed carob seeds in 1854. Now carob trees are growing wild in many areas of California and Arizona. Pasadena, California, boasts 2,000 carob trees planted along its boulevards during the depression by Seventh-Day Adventists to provide free nutritious snacks for children.

Carob contains 50 percent natural sugars, vitamins A, B1, and B2, calcium, magnesium, silicon, and iron. It is low in fat and contains some protein. Carob has medicinal uses: it alleviates diarrhea and retards emesis in babies, offers some relief of respiratory ailments, and can be a mild laxative. Carob has long been touted as a chocolate substitute, since it doesn't contain the caffeine and theobromine found in chocolate. Nor does it produce the allergic reaction some people get from chocolate. But don't expect it to taste just like chocolate. Carob has its own pleasant and distinctive taste!

Carob recipes

To make your own carob flour, first wash pods, then steam in a pressure cooker at 15 pounds pressure for 20 minutes. Cool and dry husks and remove seeds. Use a blender to grind pods into flour.

Carob powder can also be made by using a stone mortar and pestle. Crack open pods and remove seeds. Then dry husks over hot coals or leave in oven, set at lowest setting, about 6 hours. Cool and then grind by hand.

Carob Candy
    1 cup carob powder
    1 cup honey
    1 cup peanut butter
    1 cup sesame seeds, lightly toasted
    1 cup shelled sunflower seeds
    ½ cup shredded unsweetened coconut
    1 cup chopped dates, apricots, or raisins

Oil an 8-inch square pan. Heat honey and peanut butter together in a heavy saucepan. Quickly stir in carob powder and remaining ingredients. Spread in pan and refrigerate until firm. Cut into squares.

Quickie Carob Dessert
Break mature carob pods in 2-inch pieces. Serve in a pretty bowl for a fine "no-fuss" dessert. Remember, don't try to eat the seeds!

Carob Syrup
    ⅔ cup carob powder
    1 cup light mild honey
    ½ teaspoon vanilla
    ¼ teaspoon cinnamon

Heat honey in a jar sitting in a pan of water over medium heat. Mix in carob powder. Stir in vanilla and cinnamon. Remove from heat and cover jar. Store as you do honey. Serve warmed carob syrup over ice cream or yogurt.

Carob Milk

Use one tablespoon carob syrup for each cup of milk. Mix in blender for best results. Heat or serve cold.

Frozen Carob Bananas

Dip bananas in carob syrup and put on cookie tray in freezer. When frozen solid, wrap in freezer paper.

Rum Carob Cake

½ cup dates
⅔ cup rum
5 eggs
¾ cup butter
1 cup almonds
½ cup flour
1¼ cups carob powder
1 cup sugar

Soak dates in rum. Separate eggs. Cream egg yolks and butter in large mixing bowl. Grind almonds in blender and mix with flour. Mix together carob powder with sugar. Beat egg whites to stiff peaks. Alternate carob-sugar mix with almond-flour mix and blend into yolk-butter mixture. Then add date-rum mixture. Fold in egg whites last. Bake in 10-inch ring mold pan (lightly greased) or two 9-inch cake pans. Bake at 375 degrees for 25 to 30 minutes. Cake is very moist.

Chocolate:
An Aztec Legacy

*Contributed by Morgarita Mondrus Engle,
Fallbrook California*

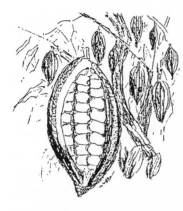

Chocolate and cocoa are both products of the cacao tree, *Theobroma cacao*. *Theobroma* is Greek for "food of the gods" and is the name given by the Swedish botanist Linnaeus after Europeans observed the Aztecs' reverence for cacao.

In pre-Columbian Mexico, wealthy Aztecs habitually drank two beverages, *cacaoqahitl* and *chocolatl*. The first was a bitter drink made by boiling cacao seeds in water; the second was a brew sweetened with honey and flavored with vanilla. The Aztecs pounded their cacao seeds with corn kernels and chili pepper to produce a mixed powder. Although the Spaniards preferred the sweet version of chocolate, they continued to mix cacao powder in with corn flour in order to counteract the high fat content of cacao seeds, which contain 50 to 57 percent oil.

Cacao seeds, often incorrectly referred to as beans, were used as currency by the Aztecs. Subjects paid tribute in cacao seeds, and large quantities were found in Montezuma's palace at the time of the 1519 conquest. Cacao seeds contain two stimulating alkaloids, theobromine and caffeine.

Cacao seeds were taken to Europe by Columbus, but chocolate did not become popular in Europe until the seventeenth century. At that time only the very wealthy could afford to patronize chocolate houses. In 1828 a process was developed in Holland to obtain powdered chocolate by pressing most of the cocoa butter out of ground, roasted seeds. In 1847 an Englishman produced eating chocolate by mixing cocoa butter, chocolate liquor, and sugar. Milk chocolate was later developed in Switzerland.

Cacao is grown only in the tropics. Although it is native to the

Americas, 75 percent of the world's crop now comes from West Africa, where disease problems are less severe. Cacao pods are botanically unusual because they are attached directly to large branches and the trunk of the tree. The harvested pods are opened by cutting or by striking two pods together. The fresh flesh around the seeds is sweet and tasty and has been compared to ambrosia.

For chocolate, seeds are removed and fermented for one week, which kills the embryo and releases enzymes needed to produce the flavor. They are dried and polished mechanically or by "dancing the cocoa" (moistening the seeds and trampling them lightly!). Cleaned seeds are heated and ground to produce an oily liquid referred to as chocolate liquor. For cocoa, it is pressed to remove much of the cocoa butter, then dried, powdered, sifted, and possibly "dutched"—treated with an alkali solution to reduce acidity and darken color.

For eating chocolate, cocoa butter is added to the liquor, which is then aerated, emulsified, and molded. Baking chocolate is almost pure chocolate liquor. Sweet chocolate is a combination of chocolate liquor, sugar, cocoa butter, and sometimes vanilla, salt, essential oils, spices, and milk. Bittersweet chocolate contains less sugar and more chocolate liquor.

Mexico's famous mole poblano, a rich sauce in which turkey, chicken, or beef tongue is simmered for hours, is a mixture of chili, onions, tomatoes, cinnamon, cloves, almonds, coriander, garlic, and bitter chocolate. Packaged mole powders ranging from mild to very hot are available in many supermarkets. Strangely enough, most traditional Mexican candies and cookies do not contain Mexico's "food of the gods."

Cacao seeds can be sprouted and grown in greenhouses or areas protected from cold and freezing weather. For a short time we had fresh cacao seed from Haiti available. Susan Caudill, my student assistant, grew them near a hot-water pipe in her West Melbourne yard. Several cacao trees are growing in protected areas in Miami and cacao trees are fruiting at the Tropical Dome in Mitchel Park, in Milwaukee, Wisconsin, and at EPCOT in Florida. Jack and I also enjoyed eating the fresh, ripe cacao fruit in Trinidad.

The Barbados Cherry

One of the greatest joys at our Florida homestead is our Barbados cherry. It is also called West Indian cherry, Jamaica cherry, and acerola. Botanically it is known as *Malpighia* species and it originated in the  West Indies. It also grows from northern South America to southern Texas.

Because of its very high vitamin C content, the Barbados cherry is used in manufacturing vitamin C tablets. You can probably find them in your local health-food store.

The Barbados cherry is actually a densely-branched bush and usually does not grow more than about 12 feet high and 10 feet wide. The deep green, shiny leaves are usually ovate, and the fruit is about 1 inch across and bright red when ripe. It is very slightly lobed and thin-skinned. It looks a little like the Surinam cherry but tastes like a Northern type crab apple to me.

Oddly enough, the 5 petal blooms on our Barbados cherry come in both pink and white. The first flowers usually appear in April, with fruit about a month later. Flowering continues all summer, and if we are lucky we have a continued supply of Barbados cherries until December.

Fruit size is improved by adequate fertilization and watering during dry periods. The Barbados cherry is said to be especially susceptible to root-knot nematodes and should be heavily mulched. We have had no insect problems to date, and after each cold spell our tree has come back and produced that season.

I've found it impossible to sprout the unusual 3-sided seeds and read that seedlings are supposed to have poor fruit. So air-layering is best for Barbados cherry plant production.

The Barbados cherry is available at some Florida nurseries. Try to find improved varieties such as B-17 and Florida Sweet. By the way, the Barbados cherry was introduced in Hawaii about 1946 by the Hawaiian Sugar Planters' Experiment Station, and the University of Hawaii has a large testing program to select well-shaped trees with extra-good fruit.

Barbados Cherry recipes

I love to eat the cherries right off the bush. Vitamin C content is 50 times that of good oranges, so 1 or 2 cherries will probably give you 75 mg, which is the recommended daily requirement.

If you have a heavy-duty fruit and vegetable juicer, use that to make fresh uncooked juice. But you can make juice by covering fruit with water, then bring to a boil and cook about 10 minutes more until soft. Strain through a jelly bag to get juice. Then run pulp through a food mill to separate it from the seeds. You can make jams and jelly, or try these recipes:

Barbados Cherry Sauce

> 1 cup Barbados cherry puree
> ½ cup sugar or honey
> 2 teaspoons lemon juice

Mix together in a saucepan and cook until glossy. Serve at once or store in refrigerator a week or two. Can also be put in sterilized jars and sealed.

Barbados Cherry Milk Shake

Blend together 1 cup Barbados cherry juice with 1 cup cold milk and ¼ teaspoon vanilla.

Dovyalis:
A Tropical Apricot

Contributed by Trude Sontille, Fort Myers, Florida

Nature makes up for not being able to grow the northern apricot in

tropical climates by giving us the dovyalis, which is also called Ceylon gooseberry. However, it tastes much like our Northern apricot.

The large shrub or small tree, originating in Ceylon, was crossed with the dovyalis from Ethiopia in Florida and resulted in the *Dovyalis hebecarpa x D. abyssinica,* giving Florida and other warm areas a tasty, small, yellow-brown fruit that will grow well in hot weather and will not freeze in temperatures above 26 degrees.

The ripe Ceylon fruit is dark purple with a velvety skin and can be used just like the cross. Both grow well in sun or partial shade and fruit from fall to spring in Florida.

The trees grow to about 20 feet and are densely branched with leaves about 4 inches long. They are vigorous and heavybearing. They like rich moist soil but do not tolerate beach areas. For larger fruit the small fruit should be thinned out. An isolated tree will produce fruit.

Dovyalis can be propagated from cuttings or air-layers. Try to graft on *D. hebecarpa* and your tree will have fewer thorns and produce more fruit.

Eat the fresh fruit, or make preserves, punches and pie fillings. Pick dovyalis when reddish-orange and refrigerate until used. The fruit is very sour and usually needs some sweetening but is much like the tasty northern apricot. This tree has given our family much pleasure. I have made many jars of the tasty jelly for gifts.

Dovyalis recipes

Dovyalis Delight

    16 dovyalis fruit (seeded)
    2 large bananas
    ¾ cup honey
    1 envelope plain gelatin
    Whipped cream or topping

Blend dovyalis, bananas, and honey together. Dissolve gelatin in a little of this mixture. Add to fruit. Pour into mold and refrigerate until set. Serve with whipped cream or topping.

Dovyalis Jelly

> 3 pounds fully ripe dovyalis
> 2 cups water
> ¼ cup lemon or lime juice
> 7½ cups sugar
> 1 pouch liquid pectin

Wash fruit and remove stems. Place in kettle with water. Bring to boil and simmer 10 minutes. Strain through jelly bag when cool. Makes 3 ¾ cups juice. Then bring juice to boil again in large kettle. Add lemon juice and sugar. Add pectin and boil 1 minute, stirring constantly. Pour into jars and seal.

The Lychee:
Mother Nature's Chinese Treat

An old Chinese saying, "When the cicadas sing, it is lychee time," is true because many lychee (or litchi) varieties ripen near the summer solstice, about June 22. Lychee fruit has been a Chinese favorite since at least 2000 B.C. And no wonder! If you taste a fresh, ripe lychee, it may become your favorite too. The juicy white fruit, obtained by removing a reddish-colored, thin, bumpy covering, tastes much like a sweet juicy grape. Lychees contain large amounts of vitamin C, as well as some calcium, iron, and phosphorus.

This member of the Sapindaceae family grows as a roundheaded tree, sometimes over 40 feet high in Florida. Its girth is usually around 30 feet. The leaves are pinnate, leathery, and shiny and have from 3 to 9 leaflets. New leaves are a beautiful reddish-copper color.

Some seasonal variations in temperature are necessary for proper fruiting. Lychees actually benefit from occasional cold snaps of 30 to 40 degrees. Flowers are on panicles in small yellowish clusters and usu-

ally develop in January and February. Lychees prefer moist acid soil with a pH of 5.5 to 7.5.

The trees do best in full sun but are not salt-tolerant. Young trees will freeze at about 26 degrees. They should be protected from wind damage and should be given adequate moisture when fruiting.

Lychees are best cultivated by air-layering. Young trees may fruit in as little as 2 years. Trees should be well mulched. A nutritional spray each spring is the only needed feeding.

Lychees are now grown commercially in Australia, New Zealand, South Africa, India, and in Florida—as well as in China. Improved varieties include Brewster, Peerless, Late Globe, and Bengal. Lychees were brought to Hawaii in 1873 and grow very well there. The Groff, a variety with a very small seed, does well there.

Several years ago we got an air-layered Brewster lychee from a Melbourne nursery. It was so interesting watching the fruit form from the blooms which occur on panicles. It held less than 1 percent of its blooms, but we had enough fruit to taste. Lychees grow well in Florida about as far north as Orlando. Some California Rare Fruit Grower friends grow lychees in protected areas.

Every year that Don Hanskutt's grove on Merritt Island, Florida, bore fruit, we would buy about 15 pounds. We froze the lychees in a large brown paper bag lined with plastic. On hot days it was refreshing to reach in and grab a lychee, peel and eat it when it was just slightly thawed.

Lychee recipes

Stuffed Lychees
Peel lychees and remove seeds. Stuff with creamed cheese, deviled ham, or softened bean dip.

Lychees With Ham
During the last 30 minutes of baking, put peeled and seeded lychees on top of ham. Delicious!

Lychees With Chicken
Arrange a frying chicken cut in pieces in a baking pan. Cover with foil. During last 30 minutes, add peeled and seeded lychees.

Tropical Fruit Salad
    1 cup peeled and seeded lychees
    1 sliced banana
    1 cup cubed papaya
    1 grapefruit peeled and segmented
    1 orange peeled and segmented
    1 cup cubed watermelon

Add your favorite salad dressing, or just use lemon juice and honey.

Dried Lychees
Use a home debydrator and follow directions. Shells become brittle and dried lychees taste much like raisins. Dried lychees are often called lychee nuts.

Love Those Loquats

One of the very best fruits for subtropical growers is the loquat. It ripens during midwinter and early spring (when few other tropical fruits are ready to eat) in Florida. It is a delightful change from citrus fruit, which are also in their prime then.

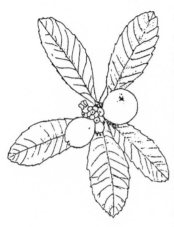

Botanically known as *Eribotrya japonica* and often called Japanese plum, the loquat actually originated in China. In Florida it grows to about 25 feet high and 20 feet wide. In Palo Alto, California, we saw a huge loquat, over 40 feet tall, loaded with golden fruit in early December.

Loquat leaves are stiff, about 1 foot long with toothed margins. Tops are dark green and undersides are coated with a whitish fuzz. The tree is evergreen and makes a beautiful addition to any landscape.

Loquats will take temperatures down to 12 degrees for short periods and are even grown outside in protected locations in London, England. Loquats will grow in a variety of soils but should be fed and watered to produce good crops. Make sure you don't over fertilize as this may cause problems. Use low nitrogen animal manures and not commercial fertilizers. They are also quite salt tolerant.

Fragrant dull-white flowers appear on long panicles mostly from October through February, but some blooming, which usually does not produce fruit, occurs during summer months. The fruit, about the size of a pigeon egg, comes in shades of yellow and orange. It is sweet with just enough acid to make it tasty.

Fire blight caused by the bacterium *Erwinia amylorvora* can be a serious problem for loquats. We lost our very first planting of four trees from fire blight, probably because we gave them 6-6-6 fertilizer.

Loquats grow well from seed, but grafting is preferred. Trees are usually available at most nurseries. Try to get the name of the variety you buy and keep a record of it. Varieties include Advance, Champagne, Early Red, Oliver, Pineapple, and Premier, which are Japanese. Tanaka and Thales are Chinese. The Wolfe Loquat (SES No. 4) named in honor of Dr. H. S. Wolfe, who planted an Advance seedling that was pollinated by an unknown parent, is probably one of the best loquats for Florida. It has large, pale yellow fruit with a delightful tart taste. The tree produces several hundred pounds of fruit each season.

In preparing loquat recipes be sure to remove the smooth, dark brown flat seeds—though in wine making, if the seed is left in the fruit you will get an "almond" flavor, which is treasured in the Orient.

Loquat recipes

Loquats can be enjoyed eaten right off the tree. Some types have quite a bit of peachlike fuzz and need to be peeled. Cut loquats in half lengthwise and fill with softened cream cheese topped with chopped nuts, ham salad, or other cheese spreads. Use sliced loquats in fruit salads. Add mashed or sieved loquats to ice cream or sherbet recipes. Many readers of my newspaper column have made and enjoyed loquat wine. See Chapter 12 for recipe.

Fresh Loquat Relish
    1 cup loquats (cut in half and seeded)
    2 or 3 calamondins (cut in quarters and seeded)
    ¼ cup honey
    1 tablespoon lemon juice
    ½ cup raisins

Put all ingredients in blender and chop only a minute or two. Put in covered glass jar and store in refrigerator. Keeps a week or two.

Martha's Loquat Pie
    3 cups loquats, seeded and sliced
    ¾ cup sugar (less if fruit is very ripe)
    2 tablespoons flour

Mix loquats, sugar and flour together. Put in unbaked 9-inch pie crust. Cover with top crust and slash for steam vents. Bake at 400 degrees for 10 minutes. Reduce to 350 degrees. Bake until crust is browned, about 35 more minutes.

Dried Loquats

Wash loquats and cut in half lengthwise; remove seeds. Place cut side down on cookie sheet. Prick each half with a fork. Put in oven set at lowest possible heat. Turn off heat after 15 minutes. Let stay in warm oven until no juice appears when you squeeze fruit between your fingers. (You may have to turn heat on again for several short periods.) You can also dry loquats in a microwave or convection oven or a dehydrator.

The Versatile Mango

Contributed by Mary Ellen Smith, Melbourne, Florida

The mango, cultivated in India for 4,000 years, is a VIF—Very Important Fruit—throughout the tropical and subtropical lowlands of the world. The mango came to the Americas in the eighteenth century and to Florida in 1833.

Mangos are versatile and a good source of vitamins A and C. You can peel and eat the fresh fruit or use it in salads and desserts. You can cook it into relishes, cakes, pies, breads, and chutney. You can dry, freeze, or preserve mangos.

The mango tree is medium to large and should be planted 20 to 30 feet apart. Seedling trees can be valuable for dooryard fruit, but trees in commercial plantings are grafted or budded. Grafted mangos bear in 3 to 4 years. The fruit ripens from May to October, with heaviest production in June and July (fruits ripen 100 to 150 days after blooming).

Mature mango trees can stand temperatures as low as 25 degrees for a few hours, with injury only to tender branches.

During the Christmas freeze of 1989 our 30-year-old mango froze to the ground. Shortly after the cold weather, we visited the ECHO research station in North Fort Myers and saw mango and avocado trees fruiting in very large pots. The intern explained, "When freezes are forecast we just wheel the trees into the greenhouse. They stay smaller in pots and are easier to pick because we don't have to climb 60-foot trees!"

Commercial plantings do best from Palm Beach and Manatee counties southward. Young trees should be fertilized every 2 months, and bearing trees 4 times a year. The mango adapts itself to many soils, but good drainage is essential. Botanically known as *Magnifera indica L.*, the mango belongs to the Anacardiaceae family. The cashew nut and poison ivy are close relatives. Avoid skin contact with mango peel. Wash your hands after handling them too! Some folks may not be able to eat mangos because of allergies.

Most commercial growers use a spray program, but many home growers do grow mangos organically. Sometimes an oil-emulsion spray will control the various scales, and a fungicide is used to control anthracnose.

Several California Rare Fruit Growers who live in warm microclimates successfully grow mangos at their homes. In Mesa, Arizona, a doctor is fruiting mangos against a north wall of his house. His problem is keeping the trees cool in 120-degree summer weather!

If you live in a cooler area try the dwarf mango Julie, which you can easily cover with quilts or move into your garage during freezing weather. It stays under 10 feet but fruit is normal size.

Mango recipes

Easy Mango Daiquiri

Put about 2 cups soft ripe mango into blender and run until pulverized. Then add 1 6-ounce can frozen limeade (or if you have ripe limes make your own). Fill can with rum and pour into blender and whirl. Add some cracked ice. Mix well and serve.

Mango Leather

Prepare about 1 quart sliced mangos. Put a small amount at one time in blender and puree. Add sugar or honey to sweeten if desired. Add 1 tablespoon lemon or lime juice. Blend until consistency of applesauce.

To prepare for food dryer, tear off a sheet of plastic wrap about 12 by 18 inches and fasten it on dryer tray with masking tape. Spread mango puree evenly ¼-inch thick. Dry until it peels off easily. This recipe can also be used for papaya, persimmon, and other tropical fruit leathers.

Sapodilla:
The Chewing-gum Tree
Contributed by Myrtle Pell Edmon, Bellflower, California

The stately evergreen sapodilla (family Sapotacae) grows to a height of 50 feet with a dense round or conical crown. It is also called sapote or

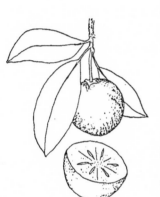

sapota and is native to tropical America, where it grows wild. It is now grown in California, Florida, and in similiar climates around the world.

Large quantities of milky latex or chicle are obtained by tapping the trunk. Southern Mexico and Central America export chicle to the United States where it is used as a base for chewing gum.

The sapodilla produces small white blooms which are followed by russet-colored, apple-shaped fruit about 3 inches in diameter. The flesh is yellow-brown, translucent, and delicious when ripe—tastes like a combination of pears and brown sugar. The fruit is primarily eaten fresh as a dessert and rarely cooked, but in Cuba and Brazil it is sometimes made into a sherbet. Avoid using immature fruit, which contains tannin and a bitter-tasting milky latex.

Although tropical in character, sapodilla does not require a high degree of humidity and if liberally irrigated can survive 27-degree temperatures. In Florida it thrives under the same care given citrus. Sapodillas prefer a rich sandy loam but will thrive on light clay or shallow, sandy soil underlaid with soft limestone. The trees are tough and pliable, which makes them more resistant to cyclones and hurricanes.

Its ability to thrive in rocky and forbidding situations on the Florida Keys is remarkable. In addition, the sapodilla has few insect or fungus enemies. The Mediterranean and Mexican fruit flies are the two most troublesome pests.

Sapodilla can be grown from seed, but for early production it should be budded like the mango. Seedlings rarely bear before 6 or 8 years. Plant seeds in flats in light sandy soil ½-inch deep. Germination takes place within a month. Pot when second leaves appear. When at least 2 feet high and visible enough to miss with your lawn mower, plant in open ground.

May is a good month for budding in southern Florida. Pick bud wood from your branches with eyes well-developed and when greenish color has a brownish tinge. Make an incision in the stock and insert the bud promptly since latex soon collects around the wound. Wrap with wax tape. After 3 weeks loosen wrap, leaving the eye exposed to start new growth.

Coleman, Pike, Wilson, and Suebelle are good varieties of white sapote to choose from nursery stock.

Sapodilla recipes

Sapodilla Rice

> 3 cups cooked rice
> 2 sapodillas, peeled and cut in small pieces
> 3 tablespoons lemon peel, chopped
> 2 tablespoons fresh ginger, chopped

Combine all ingredients and heat. Serve hot.

Sapodilla Ice Cream

> 2 cups pureed sapodilla
> ½ cup milk
> 1 cup whipped cream
> ½ cup sugar (use less if fruit is very sweet)
> 1 teaspoon lemon juice

Mix pureed sapodilla with sugar, milk, and lemon juice. Fold in whipped cream and freeze.

Indian Jujube:
Fruitful in Florida
Contributed by Bob Smith, Bonita Springs, Florida

The Indian jujube is a native of Southeast Asia, usually grown from seed. Seeds for planting should be selected from trees with good-tasting fruit since there is a large variation in both taste and quality.

The single stone contains 2 seeds which will germinate in 10 to 15  days if removed from the stone; otherwise, it will take 4 to 6 weeks to sprout.

Young trees are quite sensitive to cold. My 2-year-old tree was frozen to the ground but came back from the roots. It was also injured in later years; but now, even though the fruit may sometimes be hurt, there is only little damage to the tree. After a cold spell, the tree will usually begin new growth and new blooms.

The tree has a weeping ornamental growth pattern. Beware of the small, hooked spines at the leaf bases that may catch you as you pass. Branches weep even more when loaded with fruit. The small (about ⅛ inch) whitish blossoms are very fragrant and loved by honeybees.

The fruit looks like a small, smooth apple and turns yellowish when ripe. It reminds me of crab apples. It is still edible when brown and later becomes wrinkled like dates.

We are often asked about the difference between the Indian jujube (*Zizyphus mauritiana*) and the Chinese jujube (*Zizyphus jujube*). The trees are related and have similar growth patterns but the Indian jujube has leaves with downy lower surfaces while Chinese jujube leaves are smooth.

The Chinese jujube was introduced into South Carolina in 1837 and distributed by the U.S. Patent Office. It grows very well in dry, mild areas but sometimes does poorly in South Florida where it is usually too hot and wet. Tests conducted in Russia indicated that the Chinese jujube is very high in vitamin C and has some effect in lowering high blood pressure. It may be that the Indian jujube does the same.

Both fruit can be eaten right off the tree. In China it is sometimes boiled with millet and rice. It is also stewed or baked and candied by boiling in honey or sugar syrup. Visitors from Vietnam and Thailand are familiar with the fruit and will eat green jujubes with salt—like some Americans eat green apples. They also make a stewed dish with the fruit and I've heard it makes a very good wine. The Indian jujube has been a good producer for us in Southwest Florida and we would like to see more folks grow this tree.

The Strawberry Tree:
A Fast-Growing Delight

If you want an exotic tropical tree that will give you fruit in 2 years or less, plant a strawberry tree. Also called Capulin, Panama Berry, and Jamaica Cherry, it is botanically known as *Muntingia calabura*. It is a member of the Elaeocarpaceae family and native to tropical America.

The pinkish-red fruit, as the name suggests, resembles a strawberry. While the outside looks like a cherry the inside portion has many minute strawberrylike seeds. The beautiful white flowers looks much like a regular strawberry bloom. The taste is sweet and the odor reminds me of pink cotton candy.

The strawberry tree is an evergreen and very fast growing. You can plant the seed and harvest fruit 2 years later! If you are more impatient you can get fruit from an air-layered tree the first year. Because it grows so fast, it is a short-lived tree. Besides being a quick food producer, it will give you quick shade and privacy. It will grow to about 30 feet high and 15 feet wide. The leaves are 3 to 4 inches long and have toothed edges. Their undersurfaces are a lighter green, and they feel slightly sticky to the touch. The tree grows best in full sun but freezes at about 28 degrees. It will often recover. It likes rich moist soil and a pH of 5 to 6. It grows better in sandy soil than in limestone. It is said to have poor tolerance to salt, but I have seen fine strawberry trees growing in protected areas only a few blocks from the ocean. It has no serious pest problems, but should be pruned to prevent wind damage.

The best place to get a strawberry tree is probably from a Rare Fruit Grower meeting or plant sale. That's where we got ours.

The fruit can be used fresh in fruit salads or eaten out of hand. A half cup or more can be pureed and added to lemonade or fruit punch. In some Caribbean countries jams and tarts are made with it.

Muntingia Fruit Gelatin
    1 package (3 ounce) lemon Jell-O
    2 sliced bananas
    1 cup muntingia fruit (Remove stems and cut off blossom ends)
    1 apple, peeled and sliced
    1 lemon or lime, juiced

Prepare Jell-O using package instructions. Add lemon or lime juice. Using a single mold or 4 individual serving dishes, let cool until beginning to set. Add fruit and gently fold in. Chill until firm.

Jaboticaba:
Nature's Eye-Catching Delight

This native of Brazil is a small 15-foot tree which produces marble-size, juicy, dark purple, tasty fruit right on its branches and trunk. In full bloom the entire tree is spectacular, covered with small, delicate white flowers.

It is an attractive evergreen. The finely textured small leaves are dark green. New growth is pink tinged. Slow growing, it will take 6 years to bear fruit when grafted, and about 12 years when planted from seed. It freezes at about 24 degrees, so it can be grown in South and Central Florida. Some California growers have also reported success.

It is not salt-tolerant but thrives in rich, moist, well drained soils. It often needs added iron for good growth. (A rare fruit grower friend puts rusty iron nails under his fruit trees.) Jaboticaba will grow in full sun or partial shade.

For best results, plant at same level as in the nursery pot, water well, keep the root area covered with leaf mulch and feed with good organic fertilizers 3 times a year. Your jaboticaba tree may have 5 or 6 crops a year. It takes about a month from flowering to develop ripe fruit.

The tree may be shocked into blooming by girdling. In Miami we saw Laymond Hardy's jaboticaba tree, which he had girdled, with fruit on every branch. A Melbourne friend harvested a huge jaboticaba crop in the spring of 1990 from a tree which survived the 1989 Christmas freeze in a large tub outside.

The fruit, usually about 1 inch in diameter, has one or more small seeds. The agreeable winelike flavor is irresistible. The slightly tough skin should be removed and ground up for all recipes using the whole fruit.

Jaboticaba recipes

Jaboticaba Juice
Use methods suggested in Chapter 12.

Jaboticaba Jelly

> 5 cups jaboticaba juice
> ½ cup lemon juice
> 3¾ cups sugar

Mix in large flat-bottomed kettle. Heat to 220 degrees. Remove from heat. Skim off foam. Pour into sterilized jars and seal.

Jaboticaba Muffins

See oat bran muffin recipe in Chapter 12. Use 1 cup pureed jaboticaba.

Jaboticaba Wine

Use wine recipe in Chapter 12. (We have made jaboticaba wine. It is, without question, the very best tropical fruit wine!)

Shrubs | 6

Pyracantha

Contributed by Emily Pearce, Melbourne, Florida

The fruit of the pyracantha shrub is seldom considered food for humans. But it is edible. In fact, pyracantha is a triple-duty plant: it is attractive to birds, makes a desirable landscaping shrub, and bears fruit that can be made into delicious jelly!

Pyracantha fruit resembles miniature orange-colored apples. Like rose hips and apples, pyracantha is a member of the Rosacae family and is noted for its vitamin C content. The jelly from this fine tiny fruit first became popular in the Southwest.

The most common variety is *Pyracantha coccinea*, an evergreen shrub also called fire thorn. Six species are native from south-eastern Europe to central China.

Pyracantha has leathery alternate leaves that are minutely toothed. The white flowers are fragrant and appear in clusters (corymbs) blooming in May. The fruits are red-orange or yellow clusters of small applelike fruits (pomes) on last year's growth, lasting on the bush from November through March.

The shrub is easily grown in subtropical and temperate regions in full sun or partial shade. Although pyracantha is not particular about soil, you should avoid spraying well water directly on leaves. It will tolerate a moderate amount of salt but not dune conditions. Hosing off with a stiff spray of soapy water will usually take care of aphids or spider mites.

Why not plant some pyracantha and enjoy some edible ornamentals? After all these years of leaving them to the birds, let's try a share of the fruit and have a feast too.

Pyracantha Jelly

    3 cups pyracantha berries
    6 cups water
    Juice of 1 lemon
    Juice of 1 medium grapefruit
    1 package powdered pectin
    4½ cups sugar

Wash and stem berries. Add to water and simmer 20 minutes. Stir in lemon and grapefruit juices. Remove from heat, add water to equal 4½ cups. Return to pan, stir in pectin, bring to a boil, add sugar, and boil 2 minutes. Skim, and pour into sterilized jars and seal.

The Prickly Pear:
Nature's Thorny Treat
Contributed by Donald Ray Patterson, Titusville, Florida

Sometimes the lowliest of nature's edible plants offer the greatest feasts. Such is the case with the barbed and blossomy cactus we call the prickly pear. Her thorny armor is a mean defense, yet it seldom deters the wildlife that dines on her varied fruits.

Even man with his selective tastes can find in this wild plant a pleasant eating experience and even a means for survival. An emergency supply of water can be found in the sticky juice, which can be pressed from the insides of the stems. The fruit can be eaten raw by brushing off the spines, cutting off the ends, splitting lengthwise, and scooping out the juicy, edible pulp.

Some species bear smooth-skinned fruit that can be easily peeled and eaten. The dried seeds of the fruit can be ground into flour or used to thicken soups and stews. The young, tender pads of this cactus can be despined, sliced, and eaten boiled or roasted. They make an interesting treat toasted on skewers around the campfire and basted with your favorite barbecue or chili sauce.

Native Americans used this hardy cactus for medicine as well as food. Roasted stems were used as poultices. A diuretic tea was made from the flowers and eyewash from boiling stems. Early pioneers even boiled roots in milk and drank the liquid to treat dysentery.

The low-growing eastern variety of the prickly pear is found in Florida's coastal hammocks, sand dunes, and mangrove swamps. It also grows readily in sandy soil from New England through the South. Her taller sister, the western prickly pear, grows abundantly throughout the American Southwest and Mexico. In spring, large blossoms appear and later develop into edible fruit. A ground-hugging variety with small purple fruit grows north into British Columbia. It is even found in fields in the Carolinas as well as in central Florida.

Preparation of Fruit
Contributed by M. V.

A good way to collect the prickly pear fruit is to use a long knife and a small bucket. Hold the bucket under the fruit, cut, and let it drop into the bucket. At home, put fruit in a washtub and squirt off with a hose. This will remove most of the tiny, glasslike stickers. Use tongs to put each fruit on a cutting board. Cut off top and bottom ends. Split through top skin in center, then peel off skin. Drop peeled fruit into cooking kettle or pan. For extra caution, wear rubber gloves.

Prepared fruit can be made into juice by covering with water, cooking until soft, and then running through a jelly bag. Run the pulp through a sieve to remove the seeds. I'm sure you can think of ways to use the beautiful purple puree. You don't even have to take time to peel the fruit. An old timer in Melbourne advises, "Just put the washed fruit in a large kettle. Stick each several times with a knife. Cover with water and boil until juice is released." I tried this method but ran the juice through a paper coffee strainer just in case any of the tiny stickers were still floating around.

Eleanor Martin of Douglas, Arizona, writes:

"In Mexico cactus fruits are called Tuna, and in North Africa Barbary Fig or Indian Fig. Prickly pear, the nopal of the Aztec Indians and part of the insignia on the flag of Mexico, grows in hot, dry regions. The 'pears,' also called apples in our Sonoran desert region, are wholesome food, rich in vitamins and minerals. Their juice and jelly are a glorious shade of red!

"Though 'prickly' describes the egg-shaped fruit as well as the flat, almost round stems of the cactus, many pick it when ripe, as Indian women have done for centuries. The risk of plucking is minimized by wearing heavy gloves and using tongs.

"An elderly Mexican friend told me when she was a girl she and her companion went into the hot sunny desert to eat 'tunas' raw. They dusted off the tiny spines with a feathery weed, peeled the fruit, and spat out the many seeds. Running hands through their thick black hair removed remaining spines from their fingers. She recited a children's song for me called *Luna* (The Moon):

"Jolly moon is full
Eating cactus fruit,
Spitting rind into lake
Making strips of moonlight."

The cactus juice, too heavy by itself, has a tantalizing earthy taste. Sugar and lemon juice can be added. To make juice, the cook (still wearing gloves) washes the apples and puts them in a good-sized pot to simmer for 2 hours or more, mashing when soft. Then she lets it drip overnight through a fine muslin bag which is discarded in the morning with its load of spines, seeds, and pulp.

Young and tender prickly pear pads make a delicious vegetable called nopalitos when cut into strips or squares, boiled, and seasoned. Again, care must be taken to avoid infinitesimal thorns!

Finally, if you are over 60 and exhausted by sun and labors, revive yourself with a cactus highball. Pour sweetened juice over a glass of ice cubes and add a shot of tequila, that pure essence of cactus.

You can easily grow your own cactus pears by planting either the seed or a pad cutting. Plants start almost like magic by placing the cut portion of a pad down in sandy soil. Just keep them moist for 6 to 8 weeks or until they are rooted. In some areas, prickly pears are grown indoors in pots.

The lovely flowers, usually a shade of yellow followed by a red or deep purple egg-sized fruit, make the prickly pear an edible ornamental.

Edible fruit is found on many Opuntia species. These include:

Opuntia streptacantha (tuna cardona) is native to Mexico. It grows to 15 feet with many spines. Flowers are yellow and large. Fruit is globular and about 2 inches in diameter. Makes a fine preserve.

Opuntia rafinesquiii (devil's tongue) flowers are yellow with a red center. Fruit is about 2 inches long and 1 inch in diameter. The purple flesh is often eaten raw.

Opuntia leucotricha (duranzillo) from central Mexico grows to 10 feet. Green joints are covered with short, gray velvety hair. Flowers are deep yellow with white stamens and deep red stigma with green lobes. Fruit is whitish-yellow, round and fragrant, and is often found in Mexican markets.

Opuntia camanchica (bastard fig) is from the American Southwest where the fruit is much eaten by Indians. Pulp is sweet and juicy. Leaves are roasted.

Have you ever heard of the Mexican lemon cactus, *Ferocactus hamatacanthus?* It's used as a substitute for lemons in Nuevo Leon, Mexico. The fruit has a high acid content and Mexican cooks use it in flavoring drinks, pies, and cakes. The fruit is thin-skinned and juicy. There's also the melon cactus or Turk's-cap cactus (*Melocactus communis*) which grows in South America and the West Indies. The fruit is delicious!

Prickly Pear recipes

Cactus Fruit Jelly
Contributed by Bertha Wells, St. James City, Florida

In an 8-quart kettle put 3½ cups cactus juice, 7½ cups sugar, and ¼ cup lemon juice. Bring to a boil, stirring constantly. Stir in 1 bottle Certo (liquid pectin) and boil hard 1 full minute. Stir constantly. Remove from heat and pour into jars and seal.

I made several batches of syrup before discovering pectin is a must for cactus pear jelly. I hate using so much sugar (required with pectin), but haven't found another way to make cactus pear jelly without it.

Cactus Pad Vegetables
Contributed by Mrs. M.S. Castro, Port Charlotte, Florida

Pick only young, tender pads no more than 6 to 7 inches long. Hold over a low flame with tongs to remove stickers. The skin can also be removed since it comes off easily. Cut pads into small strips and cook in hot oil. Add finely chopped onion, green pepper, and tomato. Cook slowly. Add chopped coriander or parsley and salt and pepper to taste. One or 2 scrambled eggs can be added and allowed to cook a few minutes more. Cooked fish may also be added to make this a complete dish.

Prickly Pear Punch

> 2 cups cactus pear juice
> 1 cup crushed pineapple
> 2 cups orange juice
> ½ cup lemon juice
> 1 cup water
> 1 cup sugar or honey (less if desired)

Boil sugar or honey and water about 3 minutes. Cool. Add fruit juices and crushed pineapple. Pour over ice in pitcher or glasses to serve. (From Paul Olsen, *Peegee's Cactus Recipes,* Cactus & Succulent Society of America, Arizona.)

Prickly Pear Ice Cubes

Put prickly pear juice in ice cube trays and freeze into cubes. Keep on hand to add to fruit drinks. (From Betty Blackburn, Tucson Cactus and Botanical Society, Arizona.) Yes, cactus pears do make good wine. See Chapter 12.

The Papaya:
A Tropical Treat

When you live in the warmer parts of the world, one of your greatest pleasures is being able to pick melons from trees—papayas, of course. Well-nourished and properly ripened papayas have exquisite flavor and texture, but the neglected papaya is said to taste "like a wet dog smells."

Papayas are an excellent source of vitamins A and B. They make delightful fresh breakfast fruit but can also be canned, made into pick-

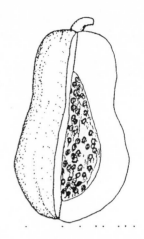

les or preserves, and even dried. Papayas can be cooked with chicken, ham, and fish, and can be used either ripe or still green. Tough cuts of beef can be wrapped in papaya leaves and made tender.

Papayas are usually grown from seed; however, they don't always produce the same quality fruit. They are improved by using seed from hand-pollinated fruit. As they come in three varieties—male, female, and bisexual—it complicates matters and requires some record keeping.

It's easiest to use the good bisexual fruiting papayas by fastening small plastic bags over the blossoms which are not removed until the ovary enlarges. For female plants, bag before flowers open and remove bag only long enough to dust male pollen on the stigma. After petals wither, pollination from unknown sources will not occur.

A major problem is finding papayas that do not grow too tall (higher than your house) and require tall ladders for picking.

Mary Noble, longtime garden editor of the *Florida Times-Union*, Jacksonville, suggests planting papaya seeds in containers any time of year you happen to find an extra tasty fruit from your own trees or from a fruit market. Prepare seeds for planting by rinsing with cold water, using a large wire strainer. Rub off gelatinous outer covering with your fingers. Rinse well. Spread thinly on trays and dry in shade. Store in airtight fruit jars with a little calcium chloride. The seeds should be viable for 2 to 3 years if properly dried.

John Burns, expert papaya grower of Sarasota, Florida, writes:

"Here's some good news for papaya lovers! A papaya hybrid, the Waminalo, has proved to be both tasty and insect-resistant.

"This three-way cross of the Florida Betty, with lines 5 and 6 of the Solo, is a bisexual and grows only 6 to 8 feet tall at my Sarasota home. It can be grown in a greenhouse or in a half barrel, which could be easily covered or moved into shelter in case of frost.

"Papayas cannot stand freezing weather! In the past I've saved fairly large papaya trees by wrapping them in fiberglass insulation.

"The Waminalo flesh is orange-yellow. The round fruit is short-necked. It is smooth and shiny, and has a star-shaped cavity. Flavor is good and flesh is thick and firm. Fruit grows to a pound or two.

"It has been grown in Hawaii about 20 years and is probably found in most supermarkets. Unfortunately, most are picked green for shipping and are almost tasteless.

"I let my homegrown fruit get soft and ripe before I pluck them. The taste is delicious! I grow my papayas from seed. I plant them in areas of enriched soil and use much organic mulch. I don't use chemicals, poisons, pesticides, or herbicides in my yard. I only spray with Maxicrop, a foliar food which contains over 62 trace minerals and vitamins. Stomata in undersides of leaves take up 8 to 10 times more nutrients than if poured on the roots. This makes my papayas healthier and the fruit tastes sweeter."

John has collected papaya seed from all over the world and has grown these varieties:

Big Shorty (6-foot trunk with sweet 5- and 6-lb fruit)
Big Sweetie (strong 8-foot tree with 5- and 6-lb fruit—from a
    Mennonite source)
Philippino Red
Brazilian
Philippino Yellow
VII—imported Florida variety, rogued out for color, shape, flavor,
    and size
Watermelon Papaya VIII—fruit grows to 10 or 12 pounds
Low growing Waminalo - grows 6 to 8 feet with very tasty, 1-2-
    pound fruit

The big problem with growing papayas is keeping fruit flies from entering the fruit and laying eggs so that when you cut open a fruit you find it filled with larvae. For several years we grew insect-free papayas because we had hard freezes that killed the insects and eggs in the ground. We also had a variety with very small center cavities and the wasp could not lay eggs in the fruit even though it stung the outside layer. But we became busy, traveled and weren't careful about picking up every single papaya that fell to the ground. Then our papayas became buggy again! Now we are excited about a 3-foot papaya from Africa which produces fruit close to the ground. It is being tested and should be available in a few years.

Papaya recipes

Papaya Juice
Place extra-ripe peeled papaya chunks in blender or food processor. Add orange or grapefruit sections or seeded calamondin halves (peel included). Add honey to taste and enough water to blend. For an extra spark add a little slice of fresh ginger. Blend again.

Papaya Meat Tenderizer
Use 4-inch squares of green papaya and cook right on meat. You can also prick roasts or fowl with fork and wrap in green papaya leaves for several hours before cooking. Prick through leaves when wrapped.

Green Papayas
Sometimes cold weather leaves you with a large supply of this green fruit. Peel, cube, and steam until tender. (I use my Chinese steamer.) Season with a little salt and pepper and a little butter. (We like to use Bohio, a blend of salt, garlic, black pepper, and oregano available at Latin markets.)

Papaya and Fruit Mix
    ½ ripe papaya, peeled and sliced
    2 ripe bananas, sliced
    2 cups chunk pineapple

Combine fruit. Squeeze juice of one lime over top.

Papaya Avocado Salad

> 1 ripe papaya
> 1 ripe avocado
> Juice of ½ lime
> 1 tablespoon red wine vinegar
> ½ cup walnut oil
> Watercress sprigs
> Salt and fresh ground pepper to taste

Peel, seed, and slice papaya. Peel and halve avocado. Place some lime juice on plate. Put avocado with cut side down on plate and rub into lime juice. Slice avocado. Fan papaya and avocado slices on chilled plate. Combine remaining juices with salt, vinegar, and walnut oil. Dip watercress in mixture. Pour dressing over papaya and avocado fan. Arrange watercress around fruit.

Guava:
God's Gift to the Tropics

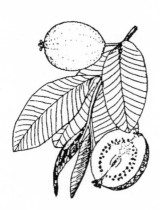

We were introduced to the guava 25 years ago when we had our first sub-tropical homestead—a 27-acre farm near Punta Gorda, Florida. It had a guava grove that had been planted to supply fruit for the local jelly factory.

I noticed then there were two main types—what I've called the "wild" fruit (about ping-pong ball sized with light yellow flesh and dark pink centers) and the "improved" (about egg-sized with either white or bright pink flesh)—besides the Cattley or strawberry guava, which is about the size of a strawberry and has a deeper pink flesh.

At first we tried to eat the fruit whole but found the hard seeds impossible to chew. Then I cut it in half, spooned out the seeds, and ran the fruit through a strainer to remove the pulp. We made sauce out of the pulp. The remaining shells could be eaten fresh or cooked.

Did you know there are more than 140 species of shrubs or small trees belonging to the *Psidium* genus? This member of the myrtle family flourishes where citrus thrives. It often grows on poor soil, and I remember finding many wild guavas growing on canal banks before they were crowded out by the Brazilian pepper. Guavas originated in tropical America but were introduced in other countries by Spanish explorers. Now they grow as far away as India and the Philippines. Varieties include Red and White Cross, Ruby, Pink Indian, Supreme, and many others.

Guava trees grow to about 25 feet but can stand much pruning and are very good for background planting. Guavas can be planted from seed. Here experts differ, with some saying the seed must be planted when removed from the fruit and others saying the seeds remain viable for a year after extraction. We've had guava seed sprout from fruit left on my work sink in the yard. Air-layering is said to sometimes produce smaller trees, which could be helpful for the gardener with limited space.

Guava trees freeze at about 29 degrees. All of ours recovered from the hard freezes of the 1980s. They can be grown in both sun and shade and even have a fair salt tolerance. They fruit mainly during July to September but often have smaller crops during the whole year.

Please don't let your guavas go to waste and let them rot under the trees. The pungent odor from rotting guavas is one you will remember! And your neighbors may complain.

Guava fruit is rich in vitamins A and C and even has thiamine, which is rarely found in fruits. The fruit also contains calcium and phosphorus, so eat guavas raw in salads and for dessert or snacks. Also use guavas in your cooking and preserving.

I like to make guava juice, which we drink fresh, and then freeze the extra for later use. It can also be canned in jars or bottles. See Chapter 12 for instructions for making juice.

Guava recipes

Guava Jelly

      3 cups guava juice

      3 cups sugar

      4 tablespoons lemon or lime juice

Measure guava juice and pour into large kettle. Bring to a boil and add sugar and lemon or lime juice. Cook rapidly, stirring frequently, to jelly stage (224 degrees) or until jelly sheets off spoon. Pour into hot sterilized jars and seal.

No-Cook Guava Jelly

Measure 3 cups guava juice. Juice 1 lemon or lime. Put juice in large kettle, add 5¼ cups sugar, and let stand for 10 minutes. Meanwhile mix 1 box commercial pectin with ¾ cup water in a saucepan. Bring to a boil for one minute, stirring constantly. Then add pectin mixture to juice mixture and stir for 3 minutes. (A few sugar crystals will remain.) Pour into sterilized jelly jars and seal. Keep at room temperature until set (sometimes takes up to 24 hours). Store in freezer, or if used within a month or two, keep in fridge.

Guava Butter

      2 cups guava pulp from jelly making

      1 cup sugar or honey

      Juice of 1 lemon or 2 limes

      ½ teaspoon ground cinnamon

      ¼ teaspoon grated fresh ginger

Run the guava pulp through a strainer to remove seeds. Add all other ingredients. Put all in large, flat-bottomed kettle. Cook, stirring often to prevent burning until mixture is thick. Then pour into hot sterilized jelly jars and seal.

Guava Mousse

> 1 cup fresh guava puree
> 1 tablespoon lemon juice
> ¾ cup sugar or honey
> 1 cup evaporated milk

To make puree, cut guavas in half, scoop out pulp, and run through a colander. Save shells for other recipes. Chill evaporated milk by placing in freezer section of your refrigerator for a short time. Pour into chilled bowl and whip until thickened. Add sugar or honey and lemon juice to puree and mix until sugar dissolves. Fold whipped milk into guava mixture and pour into freezer trays. Freeze 4 to 6 hours. Serves 6 to 8.

Cuban Guava Dessert

Put scooped-out guava shells in kettle with water to cover. Add sugar or honey to taste and simmer until shells begin to get soft. Don't overcook or they will turn into sauce. Chill. To serve, put 2 or 3 guava shells in a sauce dish and put a chunk of softened cream cheese in center.

Guava Pudding

> 4 cups guava, scooped out and cut into slices
> ½ cup orange juice
> ¼ cup lemon or lime juice
> ½ cup honey or brown sugar
> 1 tablespoon butter
> ½ cup wheat germ

Put guavas in oiled deep baking dish. Pour orange and lemon juice over fruit. Add sugar or honey and dot with butter. Bake at 375 degrees until guavas are soft. Add wheat germ and return to oven to brown. Serve hot with milk or cream. Ice cream and plain yogurt are good toppings too.

The Carissa:
A Fine Plum for the Tropics

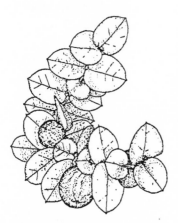

When picking the deep red fruit, take care not to get stuck by the stout thorns that grow profusely on the stems. This native of South Africa, also known as the Natal plum, has glossy, dark green leaves and white starlike flowers which sometimes have a strong jasminelike scent.

The red fruit has white milky juice—which makes one think of poisonous plants—but when heated or sugar is added, the juice turns red. Some of the carissa fruit is round; others are oval with pointed ends. The carissa, a member of the dogbane family, can be grown quite easily from cuttings or seeds. It is related to ochrosia, which has 2 scarlet fruit growing end to end and is poisonous. Don't get them mixed up. They are both used for landscaping.

Carissa grows well in many types of soils and is extremely salt-resistant. The finest carissa hedge we have ever seen is growing directly on the ocean front in Melbourne Beach.

Carissa can be grown in areas where temperatures stay above 26 degrees. It grows equally well in Florida, California, and Arizona. We saw many carissas used in landscaping parks and public buildings in Hawaii. It grows vigorously there and is often pruned into small trees and espaliers or grown into tall hedges for wind screens.

Fertilizing 4 times a year with animal manures and organic fertilizers will keep carissa producing well. This is one of the best edible ornamentals because it has no serious insect problems. The fruit is delicious picked and eaten right off the plant when fully ripe.

Varieties readily available at local nurseries include Boxwood, Beauty, Bonsai, Edulis, Fancy, Alles, and others. Some grow only a foot in height; others may grow as tall as your house.

Carissa recipes

I've heard many reports of problems getting the carissa to make a well-jellied jelly. It's best to select medium pink, not quite-ripe fruit that is still tart. If you still have problems, try this recipe:

No-Cook Carissa Jelly
    3 cups Carissa plums (1 ½ pounds)
    5 ¼ cups sugar
    1 box commercial pectin

Wash fruit and cover with water in large kettle. Cook until soft. Run through a jelly bag. Make jelly in 3-cup batches. Mix sugar with 3 cups carissa plum juice and let stand 10 minutes. Mix ¾ cup water and pectin in saucepan. Bring to a boil and boil for 1 minute, stirring constantly. Then stir into juice and sugar mixture and stir 3 minutes more. (A few sugar crystals will remain.) Pour into sterilized jelly jars. Cover with lids and keep at room temperature until set. (This sometimes takes up to 24 hours.) Then store in your freezer unless you use it in a month or two. Then keep it in your refrigerator.

Carissa Sauce
Cook fruit with sugar and water as you do cranberries. When soft, strain through colander to remove seeds.

Carissa Garnish
Cut thin rounds of fresh fruit and place them to decorate fruit salads. Or wash choice fruit. Split, remove seeds, and stuff cavities with cream cheese.

Carissa Plum Pudding
    1½ cups flour
    3 teaspoons baking powder
    ½ cup sugar
    1 cup fresh carissas, seeded and sliced

½ cup milk
1 egg, well beaten
¼ cup shortening

Sift flour with baking powder. Cream egg, sugar, and shortening. Mix with flour alternately with milk. Sprinkle fruit slices in ¼ cup flour. Add to batter. Bake in oiled pan about 45 minutes at 350 degrees. Serve with lemon sauce.

Figs:
Food for Feasting in the Subtropics

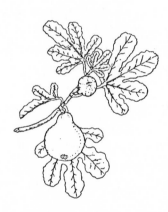

The fig, mythology says, originated from the thunderbolt of Jupiter. Widely grown in ancient Greece and Rome, figs have spread to warm areas of the world. Pliny listed 29 varieties in his day.

Figs grow best in warm, sunny areas. Soil should be alkaline with a loam cover. Rubble of brick, mortar, and rock should be added to the planting spot. These figs will remain fairly small and bear well-developed fruit. Figs are deciduous and lose their leaves with the first cold weather. But too much alkalinity may cause leaves to fall out of season. Rockledge Gardens Nursery, Rockledge, Florida, instructs folks planting figs to spread out roots and place a brick on each root before covering with soil for an increased yield.

Most amazing is that figs produce flowers inside of their fruit! This makes pollination a special project of the fig wasp (*Blastaphaga*), which enters through the tiny apex opening. It does not affect fruit when eaten.

There are four categories of figs:

1. **Caprfig**—inedible as it produces only male flowers but useful for pollen production.

2. **Smyrna**—bears only female flowers. Needs a nearby caprfig to fruit. Fruit has a superior nuty flavor.

3. **Common or Adriatic**—fruits without pollination of its female flowers. Doesn't have viable seeds.

4. **San Pedro**—has two crops, the first borne on leafless wood without pollination. The second crop will not produce mature fruit unless pollinated.

Propagation is by cuttings or layering. Layer during the summer by bending a branch down to the soil, then secure with a stake and cover with soil. By fall it should be rooted and can be detached and planted elsewhere.

Figs should be fed small amounts of mixed organic fertilizers once a month during growing season and watered well during fruiting season. Fig pests include birds, who prefer dark-colored figs. Harvest figs each morning before birds feed. Sour bugs enter open-eyed figs and destroy fruit. Pick them prior to maturity and ripen indoors.

Fig rust can be a leaf problem when the crop lingers on. Rare Fruit Growers in Miami tried removing terminals from branches, which produced an early crop before fig rust became severe.

Nematodes are the greatest problem. Planting your fig near cement walls gives better root growth conditions. Some work has been done grafting figs on *Ficus glomerata* stock. Results show they thrive in nematode-prone areas.

Other fig varieties are sold under several names. Here is a list of the most widely grown in the southern United States:

Celeste (Malta, Celestial, Blue Celeste, Little Brown Sugar)—Fruit is small, purplish-bronze, eye tightly closed. Ripens from mid-July to August.

Brown Turkey (Everbearing, Harrison, Ramsey, Lee's Perpetual, Brunswick)—Fruit is moderate size, bronze, and open eye. Ripens from July to late fall if growing conditions are good.

Green Ischia (Ischia Verte, Ischia Green, and White Ischia)—Fruit is green, which discourages birds, has closed eye. Ripens late July and August.

San Piero (Thomson, California Brown Turkey)—Fruit is very

large, purplish-black, and open-eyed. Doesn't drop at maturity but sours and splits.

Magnolia (Brunswick, Madonna)—Large bronze fruit with lop-sided appearance. Ripens mid-July to late August. The tree bears following severe freezes. This is the commercial canning fig of Texas.

Fresh figs are rich in calcium, iron, and vitamins A and C. They are low in acid and high in natural sugars. Folklore says they are good for preventing wrinkles. They are also one of my favorite fresh fruits. They may be yours too after you taste a properly ripened fresh fig.

You can pick and eat figs right off the tree. When they are properly ripened, they are a pure delight. Watch for the movie *Women in Love* based on the book by D.H. Lawrence. The hero teaches the heroine how to cut a fresh fig in quarters and eat the tiny flower blooms from each segment. So lovely. . . .

Another English writer likes his figs cut in two and soaked in an orange liqueur or sherry for an hour—and then served with sweetened whipped cream.

Fig recipes

I make a fruit salad using sliced fresh figs, cubed papaya, sliced Florida peaches, fresh grapefruit segments, seeded Surinam cherries, and the juice of a Lakeland lime. So refreshing!

Fig Chutney

In a heavy saucepan put 2 cups coarsely chopped fresh figs and 5 to 6 tablespoons sugar or honey. Mix and cook over low heat, then add:

> 6 tablespoons vinegar
> 3 small red peppers, dried and crushed
> 2 cloves minced garlic
> 1 teaspoon mustard seed
> 1 cup raisins
> ⅛ teaspoon salt

Bring to slow boil and cook about 10 minutes, stirring often to blend. Pour into jars and store covered in refrigerator. Excellent served with rice and wild game.

Fresh Fig Appetizer
> 4 cups ripe figs
> 1 cup orange juice
> 2 tablespoons sugar or honey
> 2 tablespoons lemon or lime juice

Wash figs; peel only if skin is tough. Slice figs into small pieces. Dissolve honey or sugar in orange juice and pour over figs. Chill. Serve in sherbet glasses garnished with sprigs of fresh mint.

Pickled Figs
Stick a whole clove in each fig. Cook 2 cups brown sugar, 1 cup vinegar, and 2 sticks cinnamon together until thick Add figs and cook slowly until tender. Scoop out figs and put in sterilized jars. Cover with syrup and seal. (You can add 2 ounces of rum on top of figs.)

The Surinam Cherry

Contributed by Dr. G.C. Webster, Florida Institute of Technology, Melbourne, Florida Bellflower, California

The Surinam cherry (*Eugenia uniflora*) is a bright, attractive shrub that is easy to grow in warm climates. It is a member of the Myrtacae (myrtle) family, which includes hundreds of tropical and semitropical shrubs and trees. Members of this family are found in all tropical countries around the world and occur in the Western Hemisphere, as far north as California and Florida.

The best-known relatives of the Surinam cherry are the clove tree

(*Eugenia aromatica*) of the Molucca Islands, whose dried flower buds supply cloves used in cooking, and the allspice bush (*Pimenta dioica*) of the West Indies. Other relatives are the Java plum (*Eugenia jamolana*), the Brazilian cherry (*Eugenia michelii*) the Australian brush cherry (*Eugenia paniculata*), and the famous rose apple (*Eugenia jambos*) of tropical Asia, whose delicate, yellow fruit has the faint flavor of roses.

Surinam cherries are fairly rapidly growing shrubs. The young leaves are usually brick-red and mature to a shiny green. The bushes prune readily to form good-looking hedges, and are frequently used for this purpose in Florida. If not pruned, the bushes become uneven-looking, and are not as pretty as bushes that have had even a small amount of care.

The shrubs produce small, white flowers, which are followed by bright red-lobed fruits. The edible fruits are about the size of a cherry, hence the name, but the resemblance ends there. On close inspection, the lobed nature of the fruit looks nothing like a northern cherry. I think it looks like a tiny pumpkin. The most striking difference, however, has to do with the flavor of Surinam cherries, which is distinctive and decidedly aromatic. This unique flavor is undoubtedly a reflection of the Surinam cherry's close relation to the clove and allspice. It is often a surprise to people who are expecting a cherry flavor. Surinam cherries are a good addition to the tropical or subtropical garden. They can withstand considerable neglect by the casual gardener, but will respond vigorously to a little attention. They are already well established in many parts of Florida and often used as hedges, and with their fine appearance, it is likely that they will be found in an ever-increasing number of gardens.

Surinam Cherry recipes

Surinam Cherry Preserves

Wash 2 oranges. Remove rind and then cut off white skin from rinds. Cut rinds very fine. Add 1½ cups water and ⅛ teaspoon soda. Bring to a boil and simmer, covered 20 minutes, stirring occasionally. Chop oranges, removing seeds. Add to drained orange rind. Add 2 cups cherries (heaping and seeded). Add ½ grapefruit and 1 lemon (juice and pulp of each) and 2 cups pecan halves, broken up. Simmer, covered, for 10 minutes. Measure 3 cups fruit mixture in a deep kettle. Add 5 cups sugar. Mix well. Place on high heat. Bring to a full rolling boil. Boil hard 1 minute, stirring constantly. Remove from heat and at once stir in ½ pouch Certo. Skim off any foam. Stir and skim 7 minutes while off heat. Pour into glass jars. Seal. Makes 3 pints. (From Elsie Hopes, Melbourne, Florida.)

Surinam Cherry Aspic

> 2 cups Surinam cherry juice
> ½ cup raw sugar or honey
> 1 envelope unflavored gelatin

Make juice by putting cherries in kettle; add water to cover. Cook until soft. Run through jelly bag. Soften gelatin in ½ cup cool cherry juice. Reheat ½ cup juice and dissolve raw sugar in it. Add to gelatin juice mixture and add 1 additional cup juice. Mix well. Pour into a 1-pint mold or individual molds. Chill 3 or 4 hours. Makes 4 to 6 servings.

Surinam Cherry Drink

> 2 cups Surinam cherry juice
> 6 cups water
> 1 cup orange juice
> 1 cup grapefruit juice
> 1 lemon or lime, juiced

Mix juices, sweeten to taste with honey. When you are rushed, freeze whole unpitted cherries in plastic bags or boxes. Just rinse off and freeze. Later, partly thaw, seed, and use. Add to fruit salads and punches.

Australian Brush Cherry:
An Overlooked Edible Ornamental

The Australian brush cherry (*Eugenia paniculata*), a member of the myrtle family and closely related to the Surinam cherry and guava, should be more widely grown— both for its fruit and beauty.

After more than 20 years of subtropical gardening, I was introduced to this good producer by Herb and Nellouise Kimmons of West Melbourne. Readers of my newspaper column, they asked me to identify their tree full of lovely purple cherries and planted by a previous owner.

It grows to about 25 feet in height and is upright with ascending branches and closely packed foliage. Leaves are slim and about 3 inches long. The tree is evergreen, but new growth is pinkish like the Surinam cherry. Flowers are white and about 1 inch across with many prominent stamens, which gives a very feathery look. This tree likes full sun but will grow in partial shade. It does well in many types of soil but is not salt-tolerant. The fruit, which comes in shades of lavender, has the one stone of the cherry, but rather than tasting like the northern cherry, the flavor is a mixture of guava and Java plum. The plant can be grown as a small tree or several plants can be trimmed and used as a hedge. It is propagated best from seeds but it is found at some nurseries, where it is often called the black cherry.

Our Australian brush cherry survived severe cold weather. A fine feature for subtropical fruit growers!

Both Nellouise and I have made jelly from the fruit. I made mine without commercial pectin and obtained a syrupy result. So here is her recipe using added pectin:

Australian Brush Cherry Jelly

Pick, wash, and pit 3 pounds of cherries. Put fruit in large kettle with ½ cup water and cook until soft. Strain through jelly bag. Measure 3½ cups juice. Put in large kettle and add one package commercial pectin. Let come to a boil. Add 4½ cups sugar. Bring back to full boil and boil for one minute. Remove from heat and skim off any foam. Pour into sterilized jelly jars. Makes about 5¼ cups jelly.

You can use the juice made by the above method for cooling drinks. Just add honey or sugar to taste. You can also mix the brush cherry juice with grapefruit or other fruit juices.

Bananas:
Yes, Good for the Subtropics

Ever since we've been in Florida, we've grown bananas one way or another. Since we've been at our present homestead, we've had at least one bunch of bananas each year. Folks who live in zones 9, 10, and 11 can grow bananas outdoors. We have even seen large banana plants along the highways from Destin, Florida, to Mobile, Alabama, and along the coast of Georgia.

A banana variety, Golden Rinohorn, has been found to survive minus 1 degrees when grown in Houston, Texas. Also, an old plantation banana, grown widely in Louisiana, now called TyTy Gold, also survives minus 10 degrees. A newsletter subscriber from Columbus, Georgia, resently sent me an article about bananas growing well enough to produce fruit in his inland area.

Rare Fruit Grower friends in California grow many banana varieties, but the general public doesn't seem to realize standard banana varieties can be grown for their wonderful tasting fruit in freeze-free areas. Bananas lend a tropical touch to landscaping and should be used more often around swimming pools and garden ponds.

The secret is to plant bananas where they have a continual source of water—and also some protection from cold northwest winds. Roadside canals and ditches, especially where they are protected from wind damage, are ideal places to plant banana suckers. (Suckers are the new banana "sprouts" coming from a central banana plant stalk.) Many of my neighbors have garden pools surrounded by banana plants. The pools supply the needed water.

Planting holes should be at least 3 feet wide and 2 feet deep. Fill with a mixture of compost, sand, and peat. Use a heavy layer of mulch around suckers to conserve moisture and suppress weeds.

Experts recommend heavy feeding of banana plants. This should be done monthly during warm weather. For maximum fruit production bananas should be fed ample potash. Also, if you remove all but one or two new suckers, you will have bigger bunches of bananas.

There are hundreds of banana varieties, but the Dwarf Caven dish, which grows from 5 to 7 feet, is recommended for Florida and like climates. The Apple banana, considered the finest-tasting of all bananas, grows on a thin 18-foot plant. Leaf tips are tinged with red, and the plant needs only moderate care. The Lady Finger is smaller than the Apple and has thin-skinned fruit. The leaves of the Red Jamaica banana are red-bronze-green. The fruit is large with a pinkish skin, but the plants are quite cold-sensitive.

Bananas usually need 24 consecutive frost-free months to produce fruit, which appears about 3 months after blooming. After fruiting, the stalk needs to be cut down (before it falls down) and put on your compost pile. Or if possible, allow the stalk to remain near the new plant. The stalk will nourish the plant while it decomposes.

Some growers suggest that after your plant flowers and the tiny banana hands are formed, the long tail of male flowers should be cut off about 6 inches below the last hand. This should enable the plant to produce larger bananas. The male bloom can also be peeled, wrapped in foil, and baked. A real survival food!

Banana recipes

Here is one of the best banana breads I've ever tasted:

Gladys Stellinger's Banana Bread

 2 eggs
 1 cup cooking oil
 2 cups mashed bananas
 3 cups flour (sift if it has any lumps)
 3 teaspoons baking soda
 1 teaspoon salt (optional)
 1 cup buttermilk
 1 teaspoon vanilla
 2 cups sugar (may reduce to ½ cup)
 1 cup boiling water

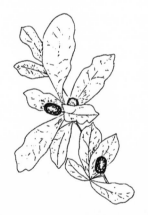

Place ingredients in bowl in the order listed above, without mixing. After pouring in boiling water, stir with spoon until well mixed. Do not use mixer or blender. Put in 2 oiled loaf pans and bake at 375 degrees for about 45 minutes.

The Miracle Fruit:
Sweetens Like Magic

Folks who attend Rare Fruit Club meetings are often treated to the amazing *Synsepalum dulcificum,* or miracle fruit, a member of the Sapotacea family. After eating one of the oval, deep red fruit, which is about the size of a coffee bean, they are offered a slice of fresh lemon or lime. Then—much to their surprise—the lemon tastes sweet!

The miracle fruit originated in Africa and was brought to the

United States by the Department of Agriculture. It is an ornamental shrub, usually branched with dense foliage that grows no higher than 6 feet and bears several crops each year. In tropical West Africa miracle fruit is often planted near native dwellings where it is used to make sour palm wines palatable, to sweeten cornbread, and to improve the flavor of stale food.

The fruit surrounds a single large seed. Eating the fruit makes the active element, glycoprotein, coat the tongue with a very thin layer. This causes sour food to taste sweet—sometimes forseveral hours.

Miracle fruit is often grown by rare fruit lovers in Florida. We find it sensitive to cold and cover it with quilts during freezes, but our plants have survived the state's recent cold years. The plant requires acid soil, and we have the best results using half barrels filled with peat moss and soil. We use acid-type azalea fertilizers but have noticed that many of the inconspicuous, white blooms fall off without producing fruit.

Miracle fruit contains the protein miraculin, which is related to the protein thaumatin found in another "sweetener" plant, Thaumatocossus. The sweetening effect of miraculin can last up to 18 hours, and plant scientists have not been able to reduce this time in order to use it as a natural sweetener for diet soft drinks. But several projects investigating possibilities for commercial use have been undertaken. The makers of Accent tried to use this fruit as a substitute artificial sweetening agent for diet drinks but failed. A project at Florida State University discovered a way to isolate the active princi-ple, but as far as we know miraculin is not being used commercially at this time.

A word of caution: One of my adult Brevard Community College students who was taking the medications Indocin and Lanoxin tasted only a small portion of a miracle fruit in my class. About four hours later he experienced a severe episode of cold sweats and irregular heart beats. In checking phytotoxin tables, no mention of miraculin or thau-matin was found. However, there are some people who are violently allergic to some proteins. So, rare fruit growers who offer samples of the miracle fruit should do so with some caution and advise folks who are on medications about possible problems.

Miracle fruit plants can be started from seed. Plants are often available at south Florida nurseries and at Rare Fruit Grower sales.

Downy Rosemyrtle:
A Beautiful Ornamental for Warm Climates

This lovely shrub with roselike blooms came from the Orient. Imported by Reasoner's Nursery in Sarasota in the 1880s, its botanical name is *Rhodomyrtus tomentosa*. It was planted at Dr. Nehrling's place at Gotha, Florida, and spread all over the surrounding woodland. It self-seeded easily in the lightly shaded, acidic, sandy soil. Now downy rosemyrtle grows wild along the west coast of Florida and inland about as far as Ocala. A member of the Myrtaceae family, the downy rosemyrtle is related to the guava, Surinam cherry, and jaboticaba.

A reader of my newspaper column, who lives in Naples, Florida, told me about the fine jelly-making qualities of the downy rosemyrtle. We visited her and sampled the jelly. Later she even sent me (via Greyhound bus) a bouquet of the downy rosemyrtle in bloom. The blooms were as lovely as a pink wild rose.

Several times we dug the plants out of deserted areas of piney woods along Florida's west coast and tried to transplant them at our homestead. Our first transplants did not survive. But we repeated the process and now have two living plants. We also picked the ripe fruit and decided they reminded us of blueberries, but were more juicy.

Recommended as a landscaping-edible ornamental, the plants may be found in some nurseries. However, in recent years environmentalists have found that in some areas of southwest Florida downy rosemyrtle has spread rampantly. They now recommend that it not be planted—which is causing some concern by Rare Fruit Growers and living-off-the-landers who use and enjoy the fruit. If you do plant the fruit, be careful and don't let it spread to your neighbor's yard!

Downy Rosemyrtle recipes

Because the downy myrtle has numerous hard seeds (like blueberries), most folks prefer to eat the fruit cooked. It can be made into juice, jams, jellies, and wine, too.

Downy Myrtle Pie

> 4 cups ripe purple downy myrtle fruit (snip off the tips
> of the calyxes)
> ⅔ cup sugar (less if fruit is very sweet)
> ¼ cup flour

Mix together and pour into an unbaked 9-inch crust. Dot with butter and cover with the top crust. Make steam vents and bake at 400 degrees for about 45 minutes.

Downy Myrtle Muffins

> 1¼ cups uncooked oatmeal
> 1 cup all-purpose flour
> ⅓ cup sugar
> 1 tablespoon baking powder
> ½ teaspoon salt
> 1 cup milk
> 1 egg
> ¼ cup cooking oil
> ½ cup downy myrtle

Combine oats, flour, sugar, baking powder and salt. Add milk, oil, and egg, mixing just enough to moisten. Gently fold in downy myrtle. Fill 12 medium muffin cups two-thirds full. Bake at 425 degrees for 20 to 25 minutes.

Monstera:
The Delicious Monster

Landscapers wanting tropical effects use the *Monstera deliciosa* both indoors and out. In frost-free areas it is hard to find a neighborhood without this luxuriant aroid climbing up a tree.

Originating in Mexico and Central America, the monstera plant has large perforated and incised leaves. The conelike fruit, 8 to 10 inches long, develop from greenish-yellow spadices with honeycomblike markings and enclosed in a waxy white spathe. Fully ripe fruit is edible and has a pleasant pineapple flavor. But until absolutely ripe the oxalate crystals in the fruit will cause throat irritation.

Monstera is often used as a ground cover in shady areas and will grow in beach areas away from salt spray. It can be started from stem-cuttings that have 2 or more buds. Plant in sandy compost, water well, and keep at 70 to 80 degrees. Seeds can be sprouted the same way. Plants are available at most nurseries and garden centers.

When used as a house plant the monstera is often subject to scales, mites, and mealybugs, but outdoors the plant seldom has insect problems. Be sure to water during dry periods, fertilize three times a year, and protect from frost.

As it actually takes 14 months from flower to fruit, waiting for your monstera to ripen is something of a problem. I suppose the conelike "roasting ears" are the most closely looked-at exotic tropical fruit in the world! We first tried to ripen ears by putting them in a glass of water and found that only a few inches of the fruit ripen each day. It is easier to put the whole fruit in a brown paper bag so it will all ripen at once.

Remove all the fruit from the core when it is fully ripe. It can then be eaten as is, or try the recipes below.

Monstera recipes

Monstera Preserves

2 cups monstera segments

Rinse fruit in colander with boiling water. Add ½ cup water and simmer in covered kettle for 10 minutes. Add 1 cup sugar and 2 tablespoons lime juice. Cook 20 minutes more. Pour into sterilized jars and seal. For room temperature storage, process in water bath.

Three Monstera Salads

#1 Combine monstera segments with apples, nuts, raisins, plain yogurt, and mayonnaise.

#2 Add monstera to diced turkey or chicken. Add celery and mayonnaise mixed with same amount yogurt.

#3 Combine monstera with grapefruit segments, orange slices, guava rings, and seeded Surinam cherries. Add lemon or lime juice and honey.

Bob and Opal Smith of Bonita Springs, Florida, showed us pictures of special packing boxes for monstera they saw in Australia and New Zealand. We predict, with the increased interest in exotic tropical fruit, folks will someday find monstera in their favorite supermarket.

Achiote:
A Natural Food Coloring
Contributed by Margarita Mondrus Engle, Fallbrook, California

When Europeans arrived in the Caribbean region, they encountered Indians who had decorated their bodies with bright red war paint. The dye they used was obtained from the fruit of *Bixa orellana*, often called lipstick tree. It is a small tree native to Central and South

America and the West Indies. A food coloring made from this fruit is still widely used in Latin America and India. The dye, referred to in Spanish as achiote or bija, is known in English as annatto.

Achiote is grown as an ornamental throughout the tropics and is becoming naturalized in many areas. The tree bears numerous pink flowers which yield heart-shaped red fruits covered with soft spines. A powdered, rust-colored food dye is extracted from the seeds and the vermilion pulp surrounding the seeds. Although the achiote fruit itself lacks flavor, dishes prepared with achiote have a distinctive, mildly pungent flavor and range from gold to bright red in color.

Achiote is sometimes referred to in Latin America as "poor man's saffron" because it is often used as a substitute for saffron in rice and chicken recipes that originated in Spain before saffron became the most expensive seasoning in the world. Saffron is the dried stigma of *Crocus sativus.* Saffron stigmas must be hand harvested, and since each plant has only 1 flower, and each flower has only 3 stigmas, 5,000 plants are required to produce one ounce of seasoning. Like saffron, achiote was once used to dye cloth, but both have now been replaced by aniline dyes. Achiote lacks the unique aroma of saffron when used as a food seasoning.

During the Middle Ages, merchants were burned at the stake when caught mixing saffron with less expensive dyes and spices such as turmeric and safflower. Today, even such traditionally saffron-dye dishes as Spanish paella are often colored with achiote. Achiote is also used for coloring margarine, butter, candy, and certain types of cheese, such as the English red Cheshire and Leicester.

Achiote is fat soluble rather than water soluble, and is one of the red dyes used in lipstick. It can be purchased as a powder called bijol, packaged in plastic bags or small cans. Whole achiote seeds are usually available in Latin specialty stores. A teaspoon or more of the powder added to the water when cooking rice, soup, or chicken produces a rich, deep golden color and a mild flavor. If whole seeds are used, they can be tied in a piece of cheesecloth or a white handkerchief, dipped in the cooking water, and removed to be dried and used again later. Seeds can be heated in oil, which takes on a red color. If they are not in a bag, they can be strained out.

The Pomegranate:
A Pleasing Fruit
Contributed by Karen Smoke, Arcadia, Florida

The pomegranate (*Puncia granatum*) or Chinese apple, is one of the most ancient fruits. Regarded since earliest times as a symbol of fertility, rejuvenation, and eternal life, it is thought to be the "apple" of the Garden of Eden.

When the new world was discovered, Spanish missionaries brought the pomegranate to Mexico and California, where it is cultivated extensively today. Pomegranates are not suited to commercial production in Florida because humidity lowers fruit quality; nonetheless it is pretty as a dooryard plant.

Two conditions caused by excessive humidity—leaf blotch, which results in premature leaf drop, and fruit spot, which reduces quality and appearance of fruit—can be controlled by periodic applications of neutral copper during the rainy season. Planting in an open, sunny area will also reduce fungal problems.

Pomegranates need some cold dormancy for best fruit production. They are hardy to about 12 degrees.

Spanish Ruby and Purple Seed are varieties recommended for Florida. Double Red is an ornamental variety, grown for its showy flowers. The Wonderful variety is cultivated commercially in California. But we have seen a Wonderful produce exceptionally heavy crops in the Melbourne area. Dwarf Red is bushy and does fairly well in Florida.

Pomegranates prefer slightly acid soil and good soil moisture. Plenty of organic amendments such as peat and compost in the planting hole and an organic mulch will help maintain soil moisture. Pomegranates require less fertilization than do most other fruit. Fertilize twice a year: after fruit harvest in November and before the spring growth flush in March. Excessive fertilization or application during fruit production causes fruit-drop, slows maturity, and reduces quality.

The pomegranate is a bushy shrub that grows to 15 feet in height. It may be trained as a tree with a single trunk if suckers are removed. But allowing it to become a shrub with multiple trunks results in earlier fruit yield and lessens the chance of cold injury.

The fruit, which follows the showy, red-orange flowers, is a large round berry with a persistent calyx and leathery skin. Inside, each of the numerous seeds is surrounded by pink-red juicy subacid pulp in a bubblelike membrane. This is the edible portion and source of the juice.

Pick pomegranates when you can hear the grains rupture when you press the skin. The fruit will last several weeks when stored at room temperature. Flavor improves with storage. Some people suck the juice off the seeds; others savor juice and seed together. The juice makes a refreshing drink, a fine wine, and can also be used for jelly or as a marinade for meat. For juice, press the grains in a hopper-type orange reamer.

How about trying to grow pomegranate? With care, it can bring a bit of paradise to your dooryard.

Fruiting Canes

7

The Youngberry:
A Good Producer for Warm Climates

Gardeners in mild regions are taking a new look at an old hybrid. Originated in 1905 by B. M. Young of Morgan City, Louisiana, who crossed a loganberry with the Austin dewberry, the youngberry is again being brought to growers' attention.

Bob Tillema, home grower and nurseryman of West Melbourne, grew the youngberry and gave me my first sample. The nickel-sized, dark purple berry is filled with sweet, subacid-flavored juice. It reminded me of the luscious boysenberry pies we tasted at Knott's Berry Farm when we lived in California. Others say the fruit tastes much like the red raspberry—

which is one-fourth of its parentage. Youngberry jelly is the color of a dark red wine, and when canned or made into juice, it needs less sugar than most other fruits.

Youngberries should be planted in good rich soil with plenty of humus and good drainage. The best planting time for Florida youngberries is from November to May. But they can be safely set out any time—if well watered. If you live in other areas, check planting dates with your local agriculture department.

Plants should be set 7 to 10 feet apart in rows, as the vines grow from 10 to 30 feet. Rows should run north and south for equal sunshine on both sides of the vines.

An unusual requirement is that the youngberry must be pruned twice in warm areas. First, just before berry ripening, prune all cane growth of the current year—except that bearing fruit.

Then, right after fruiting, all the vines that have borne fruit should be taken off trellises and cut close to the roots. The plants will be almost denuded of vines, giving growers a good chance to cultivate soil and repair trellises for the next harvest.

New vines, which become the following year's bearing cane, should lie on the ground until January and then be put up on trellises. Earlier trellising results in diminished fruit production.

Youngberries should be fed 3 times during the first year. Feed first when newly set plants put on fresh leaves. Well-rotted cow and chicken manures are recommended. The second manure application should be made when vines are tied to trellises, usually in January. The third feeding should be of a high-nitrogen fertilizer when old vines are cut off.

Youngberries should be trained on posts at least 7 feet high and set 2 feet in the ground. Three wires strung at heights of 2½, 3¾ and 4½ feet from the ground are best. Use number 12 or 14 wire.

Youngberries ripen from mid-April to July in Florida, and are ripe about a month later in California.

Remember: Youngberries bear this year's fruit on last year's vines!

Youngberry Chiffon Pie

Contributed by Mary Tillema, West Melbourne, Florida

    1½ pints youngberries
    2 egg whites, stiffly beaten
    ¾ cup sugar or corn syrup
    1 tablespoon unflavored gelatin
    ¼ cup water
    ½ cup boiling water
    1 tablespoon lemon juice
    ⅓ cup heavy cream, whipped
    1 baked 9-inch flaky pastry pie shell
    Cool Whip for topping

Wash berries and save ½ cup berries for garnish. Crush berries. Add sugar or corn syrup. Soften gelatin in cold water about 5 minutes. Then pour boiling water into gelatin mixture. Combine crushed berries and lemon juice. Cool. Fold in whipped cream and egg whites. Pour filling into pie shell and chill until firm. Use the ½ cup young-berries as garnish on top of Cool Whip to decorate pie.

Subtropical Blackberries:
Very Easy to Grow
Contributed by Margaret McCaulley, Palm Bay, Florida

During 3 years of raising cultivated blackberries, I've found them to be easy to grow, almost free of pests, and delicious tasting. Blackberries are also nourishing and low in calories. They contain vitamins A, B, and C, calcium, phosphorus, iron and even some protein. I wonder why they're not more widely grown.

My plants came mail-order as small cuttings and already a little withered. We prepared our soil with compost and hastily put in the tiny berry plants, as I was hard at work planting a spring vegetable garden. My garden had no reliable water supply and the blackberries got only an occasional bucketful, but by the following spring those that survived had formed a solid mat about 5 by 10 feet.

No record was kept of the quantities gathered during the five-week fruiting period, but during that time, the patch yielded about a quart of berries a day. I found they made wonderful jam, cobblers, and pies, and were easily canned or frozen.

We planted the Brazos, a bush blackberry. They have been quite free of pests. Mowing down the cane with your power mower and burning it after harvest may help by destroying any insects or insect eggs present. Avoid weeding by putting lots of mulch between rows. The plants send up many sprouts from the roots, which can be good or bad—depending on whether you want to increase your planting or don't want them popping up among the beans. To avoid the problem, give the plants lots of room. If you have an empty corner and would like to try blackberries, you'll be delighted with the fine berries you can enjoy with very little effort. Or check your local newspapers for pick-your-own blackberry farms. You can usually fill your tummies and your freezer on a single trip.

Blackberry Cobbler

    1 stick butter
    1 quart blackberries
    ¾ cup sugar
    3-4 tablespoons cornstarch
    1 cup self-rising flour
    2 teaspoons baking powder
    1 cup sugar
    ⅔ cup milk

Melt butter or margarine in an 8x11-inch baking pan in oven set at 400 degrees. Meanwhile, heat blackberries in kettle. Thicken with cornstarch you have mixed with cold water. Add sugar to taste. Mix flour, baking powder, sugar, and milk. Mix well and pour dough into pan with melted butter. Top with blackberry mixture. Bake at 400 degrees until dough rises to top and is brown. Make sure cobbler is done by inserting knife in center. The knife should come out clean. Serve hot or cold—good both ways!

Wild Blackberries

If you can't plant your own blackberries because you don't have enough room, you can still savor the delicious flavor. Wild blackberries grow almost everywhere around the world. There are as many as 40 different varieties in a single county.

In the subtropics, wild berries usually ripen in May and June. One year as we were driving from California to Florida, we saw children selling wild blackberries on the roadside in Louisiana. We stopped to see if we could find our own and did. In fact it took us quite a while to get home, with berry stops in Mississippi, Alabama, and north Florida.

One summer in British Columbia, we picked several kinds of blackberries in July. Later we were amazed to find the extra large Himalaya berries (developed by Luther Burbank from root stock from India and now growing wild in the Pacific Northwest) still ripening in October in the Vancouver area. And we have gone to the North Carolina mountains the first week in August for many years just because the wild blackberries ripen then.

Blackberries grow best in soil that has been disturbed, and in dug-up areas we always keep our eyes open for trailing vines. The dewberries grow close to the ground, with other types actually climbing over trees and bushes.

In the early days of our little town, every vacant lot supported a blackberry patch. One of our very best finds was a blackberry patch in a wrecking lot in Winter Garden, Florida.

Make sure that you identify wild berries correctly. The fruit of the poisonous *Lantana* looks slightly like a blackberry. Remember that blackberries have S-petaled white blooms. It takes an experienced botanist to identify some of the *Rubus* species. And when you go blackberry picking, wear clothing to protect your body from thorns on the vines—and dust some sulfur powder in your shoes to discourage chiggers.

Mysore Raspberries:
A Black Raspberry for Warm Climates

Contributed by Laymond Hardy, Miami, Florida

When grown with adequate soil, moisture, and nutrients, the black Mysore or tropical raspberry (*Rubus albescens*) is a very productive and delicious fruit in areas where killing frosts are rare. However, cool weather does aid in inducing heavier fruit set. It is most likely named after Mysore, a city in India, where it probably originated.

During cooler winters, good fruitset should be about 4 months, lasting from March through June. Good fruiting can be expected from healthy year-old plants, but optimum bearing usually begins the second year. A 70-foot row of well-established 2-year-old plants should yield about ½ gallon of berries every third day at the season's peak.

Propagation is usually done by tip-layering, or by burying the entire outer 4 inches of a vigorously growing terminal shoot in a pot of soil. By the end of 3 weeks a strong root system should be well established in the pot and a young "upside-down" plant can be severed from the parent plant. The "upside down" plant will rapidly reorient itself and will be ready to plant in the open ground. Eight plants spaced about 5 feet apart should provide all the berries needed for the average family—plus local bird populations!

Water young plants growing in nursery pots (and for a while when planted in their permanent field rows) once a week with a water-soluble fertilizer like Nutrisol. A follow-up of standard all-purpose fertilizer, like 6-6-6 with additional magnesium oxide and about 40 percent organic nitrogen, applied once a month has given good results in south Florida rocky soil. Well-established plants thrive on well-drained but moist organic soil.

Mysore raspberry plants should be placed 5 feet apart in rows 12 feet apart. Vines are best supported by wire trellises. Use 2 wires, one above the other stretched between posts. The lower wire should be 30 inches above ground with the upper wire 60 inches above ground. Tie young vines to trellises with coarse manila twine.

Leaves turning yellow can be caused by too high a water table or from a deficiency of iron, manganese, magnesium, or zinc. Each deficiency has its own distinct yellow color pattern. Three applications of good nutritional spray applied at 2-week intervals should prevent this problem for a year or more. Bees like to work Mysore raspberry flowers, but the plants do not require pollination to set fruit and produce viable seed. The young plants produced from Mysore raspberry seeds are not true sexually produced seedlings, but nucellar embryos that come true to type as do other plants of a nonsexual origin, such as bulbs, cuttings, or grafts.

Mysore Raspberry Recipes

Eat the berries fresh, or use them to make milk shakes, cobblers or pies, raspberry jelly, or a topping for ice cream.

Raspberry Milk Shake
Using a 5-cup blender, churn:
    1½ cups raspberries
    1 to 2 tablespoons sugar or honey
    2 cups milk

Blend until smooth. Add enough milk to fill churn. Stir with spoon. Let stand for 5 minutes. All seeds will settle to bottom and can be saved for planting. They will be perfectly clean after rinsing with water.

Raspberry Jell-O

For 1 package of raspberry or black cherry Jell-O, use about 2 cups Mysore raspberries. Follow package directions. Cobblers and pies can be made as you would with other berries. Mysore raspberry flavor is best enjoyed with vanilla ice cream.

If the berry plants are kept well drained but never left to dry out, there should be large quantities of fruit.

Grow Great Roses in Warm Climates

Contributed by Tom MacCubbin, Urban Horticulturist, Orange County Extension Service, garden writer, radio and television host, Orlando, Florida

Roses require just a little time. About an hour of care a week is all that's needed to produce long-stemmed blossoms from bushes that grow more than head-high.

"It's as easy to care for 10 bushes as for 1 or 2," says Mark Nelson, vice president of Nelson's Florida Roses, an Apopka, Florida, company that was started by his grandfather 30 years ago. The company produces more than 100 varieties for market and trial in the Sunshine State.

According to Nelson, roses have an unwarranted reputation for being hard to care for. One reason is that not all roses grow well in Florida and other warm areas. Even some of the All American Selections are not right for the subtropics. The climate is hot and pests are plentiful. Home growers need varieties that produce well in sandy soils.

To keep a supply of good, vigorous bloomers, Nelson is bringing back lots of nonpatented varieties, including Christian Dior, Mothersday, Mister Lincoln, Queen Elizabeth, and Tropicana. Many are more than 17 years old but are time-tested plants that have sturdy stems and maintain green foliage.

Helping to keep subtropical roses vigorous is the rootstock on which the varietal roses are grafted. The Nelson Company roses are grafted on the Fortuniana rootstock. Another rootstock used in this climate is Dr. Huey. Nelson finds that roses grafted onto Fortuniana grow vigorously all year and are nematode-resistant. At Nelson Roses, each year's All-American Rose selections are being grafted onto Fortuniana stock for fall planting.

To grow roses, start with a planting hole twice as big as the root ball of the purchased plant. Mix peat and rose fertilizer with the fill soil and some lime if a soil test shows that it is needed. Then plant the rose at the same depth as it was in the container. A mulch will be needed. Two inches of pine needles are ideal because they resist compaction, but more-available mulches of compost and pine bark are also satisfactory.

Rosarian Charles Alcott of Mount Dora, Florida, does even more for his prize-winning roses. He recommends planting in a mix of equal parts of composted bark, peat moss, and compost or improved soil. To this the gardener must add a phosphorus source such as Mag AmP (a 7-40-6 commercial fertilizer), bone meal, or superphosphate at labeled rates. If you want the "full treatment," Alcott also suggests mixing in alfalfa pellets, vermiculite, perlite, or kitty litter. Let this "super soil" set for a few weeks before setting in plants.

Roses are heavy feeders and respond with prolific growth and big blooms to frequent fertilization. Nelson suggests using a general mix such as a 6-6-6 analysis at 2½-pound rate spread over every 100 square feet of bed. Be sure to broadcast the product under the entire plant—don't just ring the plant with fertilizer.

Some additional weekly care will keep your bushes full of foliage and blooms. If necessary, treat plants for mites and blackspot fungus. During hot, rain-free weather, water twice a week, applying three-fourths of an inch of moisture at each irrigation. When the weather cools, once-a-week watering is usually adequate.

A dentist in Glendale, Arizona, grows hundreds of roses successfully in the desert there, using a large amount of mulch and drip irrigation. However, his special secret is preparation of his planting holes. He digs them 18 inches wide and 18 inches deep, placing the soil to one side. He fills half the hole with mulch and adds 1 cup of triple phosphate and soil sulfur, blending it with new soil cut from the sides and bottom until the hole is 2½ feet wide and 3 feet deep. This is much easier than mixing it in a wheelbarrow.

Using Roses

Roses have always been very special, and over the years many foods, cosmetics, and even medicines have been made with them. For these purposes, you may use any type of rose. However, use a rose variey for your climate.

Rose recipes
Contributed by Betty Mackey, garden writer, Longwood, Florida

Rose Potpourri
Spread out 6 cups of fresh rose petals to dry on a screen, either in the sun or in an oven set on the lowest setting. They should become dry enough to crumble, but still retain fragrance. (Drying outdoors may take 2 or 3 days. Bring them inside at night so they will not be wet with dew.) When completely dry, crumble them a little, and mix with 2 tablespoons dried minced citrus peel, 1 crushed cinnamon stick, 2 crushed bay leaves, and 1 tablespoon minced orris root (available at pharmacies). Orris serves as a fixative. Pack loosely and seal the mixture in jars. Shake from time to time. After 3 weeks, your potpourri is ready to use in little bags or pillows as sachets, or in dainty boxes with removable covers.

Rose Tisane

¼ cup minced fresh rose hips (seeds of the roses left
  after petals fall off)
4 cups boiling water

Place minced rose hips in a large china teapot. Pour in 4 cups of rapidly boiling water. Steep for 6 minutes. Strain as you serve. Sweeten with honey. Makes 4 or 5 servings.

Rose Hip Jam

Collect rose hips after they are fully ripe. Do not use hips from bushes that have been sprayed with insecticide. Gather a pound of rose hips (approximately 3 cups). Boil them with 1 cup of water, then simmer until the fruit is very tender. This may take half an hour or so. Put through a strainer or food mill. Mix the pulp with 1 pound of sugar. Simmer until thick. Pack in sterilized jelly jars and seal with paraffin. This jam is very tangy and can be served with meat and game or on toast.

Plants with Edible Roots and Bulbs

8

Chinese Water Chestnuts

Contributed by Margaret Normile,
Melbourne, Florida

Chinese water chestnuts, a standard ingredient in most Chinese dishes, have been grown successfully by my family on a plot of land west of Melbourne. With a little planning and preparation, they may be grown in a backyard plot.

This unusual vegetable bears a superficial resemblance to the corm of a gladiolus. Beneath the smooth, chestnut-brown skin is firm, white flesh like that of an apple. It has a nutlike flavor. Homegrown Chinese water chestnuts and a supply of fresh bean sprouts from your kitchen are the basic ingredients for chow mein and chop suey. They are also delicious in American stews, casseroles, and salads.

The plant is a sedge, similar to rice, and grows in water. It lacks leaves. Instead, numerous upright tubular stems grow to a height of 3 to 5 feet. The corms are dug from the ground, pared, and sliced about ⅛-inch thick for cooking. Unlike most vegetables, Chinese water chestnuts retain their crisp texture when cooked, adding excellent quality to any dish.

Planting should be made in level, fertile areas surrounded by a low dike that can retain 4 or 5 inches of water throughout the growing season of 7 months. The corms are planted in rows 30 inches apart and 4 or 5 inches deep. The area is flooded for 24 hours, then drained. When the plants are 8 to 10 inches high, the ground is flooded for the remainder of the growing season. The plants grow fast, spreading out from the mother plant.

Chinese water chestnuts require large amounts of plant food, but even the most favorable soils must be tested for acidity, as the plant will not do well in acid soil. Ground limestone in the soil or broadcast on top of the water is one way to take care of this problem.

The water is drained from the plot about a month before harvesting. Draining hastens the maturing, and makes the corms easier to harvest. Production may run as high as 1 pound per square foot.

Suggestions for the Backyard Gardener

Walter O. Hawley, researcher at the U.S. Plant Introduction Station in Savannah, Georgia, suggests amateur growers use a washtub or a wood form lined with polyethylene to make a watertight base. It should be about 10 inches deep. Either soil or a planting medium can be used. Just 2 or 3 seed corms should produce from 20 to 60 pounds of water chestnuts.

Plants need about 7 months of frost-free weather to mature. It's best to begin in March. Start by sprouting 3 fresh water chestnuts from the produce departments of your local store.

Cut a drainage hole in your mini-pool. Make sure to get a plug, as you need to keep water in for the major part of the growing season. Don't forget the lime. We added a shovelful of gravel and covered it with about 7 inches of a good rich soil mix, builder's sand, manures, and bonemeal. Use your pH tester to make sure the soil is alkaline (over 7). When shoots are about 8 inches high, they are just right for planting. Arrange them in the pool and cover with water to settle the soil around the corms, then drain. When sprouts are 10 inches high, flood pool to top. Feed with manure tea (manure soaked in water), but be careful not to let the soil pH become acid. When mother plants send out daughters, drain pool and add cow-manure slurry. Drain pool, add slurry, and leave overnight. Reflood the pool next day and watch the plants grow. By late summer, the pool will be full of reeds. When you see insignificant flowers, add more slurry to feed the heavily developing corms. About November 1, drain the pool and let dry for about 4 weeks. Then dig out corms by sifting through a screen. Mature corms are chestnut brown; immature corms are still white. Put corms in shallow trays to dry slowly in the shade. Then store in glass jars in refrigerator. Peel when ready to use. It is laborious, but the delicious fresh taste makes it worthwhile.

Jicama:
A Root Crop for the Subtropics

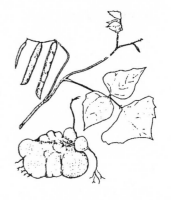

Jicama is a member of the Legumonosae family and is botanically known as *Pachyrhizus erosus,* though it is often called yam bean. The vine spreads prostrate over the ground with the slender stems reaching 10 to 15 feet in a single season. Jicama is now produced commercially in Mexico, California, Arizona, and on an experimental basis in Florida, although cultivation patterns vary. In some cases seeds are just broadcast into plowed soil and tamped in; others are planted in raised beds 2 by 100 feet, with seeds placed at 15-inch intervals.

The younger and immature pods are sometimes cooked and eaten like string beans, but the seed pods are said to be somewhat poisonous unless the epidermal hairs are removed before cooking, which is done by rubbing pods in sand. Mature pods have a high rotenone content— enough to make an insecticide when powdered. To be safe, all jicama bean pods should not be eaten and should be kept away from children and pets who could chew on them! The flesh of the tuber is white and mostly carbohydrate. The juice is either watery or milky, with the Jicama de Agua variety most commonly cultivated. Three of the best species for home production are *P. tuberosus Spreng, P. erosus Urban,* and *P. ahipa.*

Dr. C. A. Schroeder, professor of biology at the University of California, Los Angeles, wrote, "We had rather good results when jicama seeds were soaked overnight, planted in a flat, and then transferred to small pots for a short time before planting in the field. Spacing of approximately 15-18 inches in the row has provided good results."

We have grown jicama successfully in our Florida gardens, and I especially enjoy the blue sweet-pealike flowers. The beans look much like regular green beans, so again, make sure you warn family and friends not to eat them. Only use the turniplike tuber!

Jicama recipes

Jicama Fresca (Jicama Appetizer)
> 1 to 2 pounds jicama, peeled and sliced ¼ to ½-inch
> thick
> 1 teaspoon salt
> ¼ teaspoon chili powder
> 1 lime, cut in wedges

Blend salt with chili powder and put in a small bowl. Arrange jicama slices on serving tray with bowl of seasoned salt and lime wedges. Rub lime over jicama, dip in salt, eat and enjoy. If you are trying to eliminate salt from your diet, try just dipping the jicama slices in lime juice and hot pepper sauce.

Jicama Salad

> 2 cups peeled and diced raw jicama
> 1 green bell pepper, seeded and diced
> 1 cup diced cucumber
> ¼ cup olive oil
> 2 tablespoons white vinegar
> ½ teaspoon crumbled oregano
> Salt and pepper to taste

Combine jicama and other vegetables. Pour olive oil and vinegar over the mixture and add oregano. Add salt and pepper if desired. Serve chilled.

The Mexican pavilion at EPCOT serves a tasty salad combining sliced jicama, onions, and oranges.

Ginger:
A Lively Spice
Contributed by Kaye Cude, herb expert, Fort Myers, Florida

Ginger (*Zingiber officinale*) is a member of the Zingiberaceae family and is related to cardamom and tumeric. It is considered a native of India and China and has been used in medicine, food preservation, and food flavoring for many years. Spice traders in Basel, Switzerland, sold their wares during the Middle Ages on a street named Imbergasse, or "Ginger Alley." The price of a pound of ginger was then the same as the cost of one sheep. By 1500, ginger plantings were established in the East Indies, and several islands, including Jamaica, exported it. The ginger rhizome, sometimes called "hand" is harvested and used as ginger root. The long white roots from which the rhizomes grow require rich, moist organic soil at least 9 inches deep.

Ginger grows to 4 feet in height. Its fragrant, narrow leaves are arranged alternately and emerge from a sheath that enfolds reedlike stems. Ginger seldom blooms and when it does the small bloom is so insignificant it often goes unnoticed. The plant is dormant a part of each year.

To grow your own ginger, cut a plump rhizome with a strong ginger fragrance into 1½-inch pieces. Place in a shallow tray of water until 1-inch sprouts appear. Press sprouted rhizomes into soil and cover with light moist mulch 2 inches deep. Use a balanced fertilizer with minor elements. Place small pieces of concrete block on top of the mulch to supply needed lime. About 3 months after planting, small pieces of ginger can usually be  cut without endangering the plant. Before the leaves die back (9 -11 months), the entire crop should be harvested with a potato rake or fork. (We have been too busy to harvest all our ginger at once and found it keeps producing. Nice to be able to go out and dig a little fresh ginger when needed! M.V.)

Never store ginger rhizomes in a dark container without moisture or dehydration will be a problem. Commercially dried ginger is treated by various methods before the drying process. Small pieces kept for replanting are best stored in cool, shady locations. Always keep your best clones for propagation.

Ginger recipes

Add 1 teaspoon freshly grated ginger to your favorite yogurt salad dressing. Add ½ teaspoon dried ginger to 3 to 4 cups fruit-leather puree for a more spirited "roll-up." Leaves of the ginger stalk may be used fresh or dried, in much the same manner as bay leaves. The fresh stalk is a beautiful addition to many flower arrangements. Grate ginger with the grating blade of your food processor; chop with the metal blade. It is seldom necessary to peel ginger rhizomes; just wash thoroughly.

Recent studies have found some toxic qualities in ginger (also mustard, sassafras, and licorice). Don't overdo the use of any of these items at one time.

Ginger-Papaya Marmalade

Thinly slice 2 lemons and cook in 2 cups water for about 30 minutes. Boil 1 teaspoon freshly grated ginger root, 3 cups honey, and 1 cup water to make a syrup. Combine lemon and syrup mixture with 8 cups firm, ripe papaya chunks and cook slowly for 30 minutes or until thickened. Don't let it burn! Pour into hot, sterilized jars and store in refrigerator. One-fourth teaspoon ground ginger can be substituted, but flavor will be slightly different.

Tropical Rice

Saute 1 cup thinly sliced onions in 3 tablespoons safflower oil until onions are transparent. Add 1 tablespoon grated orange rind, ½ teaspoon ground ginger (1 tablespoon freshly grated ginger can be substituted), ¼ teaspoon dried thyme (or 1 teaspoon fresh thyme), ½ cup orange juice and 2 cups chicken stock (or 2 cups chicken bouillon). Bring to boil. Add 1 cup long grain rice to boiling mixture and stir once with a fork. Cover and simmer over lowest heat for 25 minutes or until all liquid has been absorbed. Stir again before serving. Garnish with 1 cup chopped fresh parsley. Good with chicken, turkey, or pork.

Chinese Ginger Vegetables

    12 whole button mushrooms
    12 whole snow peas
    1 stalk celery, sliced thin
    1 green pepper, sliced
    ¼ cup sliced water chestnuts
    1 cup broccoli, cut up
    2 stalks bok choy, sliced
    1 carrot, thinly sliced
    1 cup cauliflower, cut up
    1 cup sliced zucchini squash
    ¼ cup soy sauce
    2 tablespoons white wine
    2 tablespoons peanut oil
    ¼ cup water

> 1 teaspoon fresh ginger, grated
> 1 clove garlic, crushed
> 1 tablespoon arrowroot (available at health food stores.)

Dissolve arrowroot in ¼ cup water and set aside. In skillet (or wok) stir fry mushrooms and vegetables in peanut oil until glossy. Combine soy sauce and wine and dissolve spices in mixture. Add to vegetables and stir fry about 10 seconds. Cover and steam about 1 minute. Remove lid, pull vegetables to one side, and add dissolved arrowroot to liquid in pan. Cook for another minute to thicken. Mix with vegetables. Serve with rice or cooked noodles.

Jerusalem Artichokes:
A Crisp Treat

Contribution by Betty Mackey, former associate editor of Living off the Land, *subtropic Newsletter, now living in Wayne, Pennsylvania*

You can find them or you can grow them. Either way, Jerusalem artichokes (*Helianthus tuberosus)* are plants with hidden values. Closely related to and resembling sunflowers, their hidden offer is a large clump of deliciously edible tubers growing underground below the main stem. They resemble fresh ginger root.

Jerusalem artichokes, which have been cultivated by Native Americans, occur in the wild from the central Eastern seaboard down to Florida and, at the same latitudes, westward to Texas and north to British Columbia, but they can be cultivated in all parts of the U.S. mainland. They are a persistent plant often found in abundance on old farmsteads and homesteads. They are sold in many fresh vegetable markets during fall and winter.

Sometimes they are called sunchokes to distinguish them from globe artichokes, which are members of the thistle family. Jerusalem artichokes are easy to spot while in bloom. Healthy specimens reach a height of 8 or 10 feet. The 3-inch, sunflowerlike blossoms are golden with greenish centers.

Make note of where you find the blossoms so you can return 2 months later to dig the tubers. The flowers are lovely cut for indoor vases. Cutting them helps the plants direct their energy to the fast growing tubers.

Because Jerusalem artichokes are rarely found growing wild in Florida, some of their fans don't realize that they can grow this traditional favorite in a subtropical home garden. Tubers, the larger the better, are planted 8 inches deep and 12 inches apart. Choose a sunny, well-drained spot and enrich it with manure and compost. Jerusalem artichokes won't put up with plain sand. Water enough to keep the soil moist, but don't let it stay soggy.

Plant tubers in fall or winter for an easy crop 130 days after they sprout. When the plants are over 5 feet high and near blooming size, they topple over easily, so stake them for support. Each plant yields several pounds. Though the plants are vigorous and hardy, the tubers sometimes rot during a hot, wet summer. Let them grow during the cooler part of the year and then harvest before summer. Test for ripeness, then dig, wash, and refrigerate the tubers in plastic bags with ventilation holes. Refrigerated, they remain fresh a long time. I once kept some nearly a year and then had them grow well when planted out. In the coldest part of the year, ripe tubers can remain underground until you wish to harvest them, but keep an eye on them because they begin to sprout even in weather that we consider cold. If picked while green, the tops make good silage for cattle.

Jerusalem artichokes may be eaten raw or cooked as soon as the tubers swell. Though they are unrelated to globe artichokes and look nothing like them, the flavor is similar. One unusual feature is that their texture and calorie content change as they ripen. They are high in insulin, which converts to sugars in storage, causing them to progress from 22 to 235 calories per pound. They also provide protein, calcium, phosphorus, iron, and vitamins, with only a trace of fat. They may be used in the same ways as waterchestnuts and potatoes, and then some! (Note: Jerusalem artichokes have the same side effect as cucumbers have on some folks, i.e., they can produce gas.)

Jerusalem Artichoke recipes

Like globe artichoke hearts, Jerusalem artichokes taste wonderful sautéed in butter, mixed into salads, combined in casseroles with meats or cooked vegetables, or even pickled. They cook faster than potatoes, but can be eaten raw.

Wash your Jerusalem artichoke harvest with a vegetable brush under cold running water. Pat dry with towels and refrigerate in ventilated plastic bags until needed. Then just rinse, peel, and slice for recipes like these:

Betty's Szechuan-Style Chicken and Artichokes with Orange Peel

> 3 boneless chicken breasts
> 1 orange
> 1 cup peeled and sliced Jerusalem artichokes
> 1 onion
> 6 cups thinly sliced bok choy or Chinese cabbage
> 2 small hot red peppers
> 3 tablespoons cooking oil
> 3 tablespoons soy sauce
> ¼ cup vinegar
> ½ cup orange juice
> ¼ cup water
> 2 tablespoons cornstarch
> 2 tablespoons sugar

Prepare rice. Make sauce mix by combining cornstarch and water in bowl. Stir in vinegar, sugar, and 1 tablespoon soy sauce. Set aside. With a sharp knife shave off 1-inch circles of orange peel, taking as little white as possible, and set aside. Add orange juice to sauce mix.

Cut chicken into bite-size slices and marinate in 2 tablespoons soy sauce. Set aside sliced artichokes, onions, and bok choy together.

Discard seeds and slice red peppers; set aside. Set large skillet or wok on high heat. Add oil. When hot, saute orange peel circles and hot peppers until edges blacken, remove with slotted spoon (powerful aroma!). Stir fry chicken 1 minute. Add sliced artichokes, bok choy, and onions; stir fry 1 minute. Return orange peel and peppers. Stir and add sauce mix, stirring until sauce thickens (2 minutes or so). Serve over rice. Serves 4.

H.P. and Mrs. Bradley, a Merritt Island, Florida, retired couple who grew and used Jerusalem artichokes, invited me to visit. They presented me with two jars of their homemade artichoke relish, his and hers. (M.V.)

Mrs. H.P. Bradley's Artichoke Relish

    1 gallon chopped artichokes
    1 quart chopped bell peppers
    1 quart chopped onions
    2 quarts vinegar
    5 cups sugar (or less)
    2 tablespoons each of celery seed, mustard seed,
        tumeric, and dry mustard
    1 cup plain flour with enough water to make into
        medium paste
    1 cup salt

Soak chopped artichokes and onions for 2 hours in salt water using 1 cup salt and water to cover vegetables. Drain well. Bring sugar, vinegar, and spices to a boil. Add artichokes, onions, and peppers. Hard boil for 5 minutes. Add flour paste and boil for 5 more minutes. Pour into hot, sterilized jars making sure to cover tops of artichokes with juice. Seal. Makes 14 pints.

Mr. H. P. Bradley's Artichoke Relish

Wash and scrub artichokes. Then make brine with 3 parts vinegar to 1 part water. Add about 1 teaspoon sugar, salt to taste, and spices including mustard seed, celery seed, and whole black pepper. Bring to a boil. Put artichokes in hot jars. Pour brine over and seal.

Sweet Potato:
The Lazy Man's Vegetable
Contributed by John Burns, Sarasota, Florida

The sweet potato is frost-hardy in the tropics and subtropics. It usually needs no watering, feeding, or weeding. Roots weigh from 1 to 7 pounds. They are delicious as well as nutritious. They can be baked, stewed, steamed, fried, or even eaten raw. Even the leaves are edible. In China and the Philippines they are stir fried and called "poor man's salad."

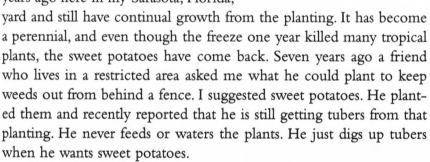

Botanically known as *Ipomoea batatas,* the sweet potato originated in South America. It spread to Micronesia and the Orient many years before Columbus took it back to Europe. There are hundreds of sweet potato varieties, including those with yellow, white, and purple flesh.

I planted sweet potatoes over 10 years ago here in my Sarasota, Florida, yard and still have continual growth from the planting. It has become a perennial, and even though the freeze one year killed many tropical plants, the sweet potatoes have come back. Seven years ago a friend who lives in a restricted area asked me what he could plant to keep weeds out from behind a fence. I suggested sweet potatoes. He planted them and recently reported that he is still getting tubers from that planting. He never feeds or waters the plants. He just digs up tubers when he wants sweet potatoes.

After digging, most sweet potatoes start to decompose or get weevils. I usually just dig out the amount of sweet potatoes I need and keep the rest stored in the ground. Sweet potatoes make a fine ground cover. When planted around ditches they help prevent erosion. If you want to start sweet potatoes in your regular garden, consider planting in raised beds that have been prepared with fertilizers, manure, and organic materials. Cuttings should be inserted about 15 inches apart, buried for half of their length. If planted in dry times, cuttings should be protected from the sun using palm fronds or grasses. Crops usually take from 2 to 5 months to mature, depending on variety and growing season. Potash is vital for good production, so use your fireplace ashes in planting beds. Sweet potatoes are rich in vitamin A. They also contain calcium, phosphorus, iron, sodium, and niacin. They are rich in carbohydrates, and have some protein and a very small amount of fat. A large baked sweet potato usually contains about 150 calories. A variety called Vardaman outyields most other sweet potatoes. It has a deep orange color and very tasty flavor.

(We followed John's advice and had low-care sweet potatoes growing in our garden for 5 years. However, our potatoes recently have been bothered by sweet potato weevils. We must plant them in new soil and sterilize our old patch.—M.V.)

Sweet Potato recipes

Most sweet potato recipes call for additional sugar, sweet syrup, and even marshmallows. We think sweet potatoes should be used in their "natural state" with only a very little added sweetening or ideally none at all.

Sweet Potato Salad
1½ pounds sweet potatoes
1 cup thinly sliced celery
½ cup thinly sliced green onion
1 large red apple, diced
½ cup mayonnaise
1 teaspoon mustard
½ teaspoon grated orange peel
1 tablespoon orange juice
2 tablespoons finely chopped, crystallized or

fresh ginger

Cook sweet potatoes in water to cover until fork-tender. Drain and cool. Peel and cut into ½-inch cubes. Combine sweet potatoes, celery, onion, and apple.

Mix together mayonnaise, mustard, orange peel, juice, and ginger. Pour over sweet potato mixture. Add salt and pepper to taste. Cover and chill for at least 2 hours. Serve on a platter covered with crisp greens.

Sweet Potato Pone Pie

1½ cups cooked sweet potatoes, mashed
½ cup brown sugar (optional)
¼ teaspoon cinnamon
2 eggs, beaten
1 tablespoon lemon juice
1 cup milk
1 unbaked 9-inch pastry shell
Pecan halves

Combine first 6 ingredients. Beat until smooth. Pour into pastry shell. Bake at 450 degrees for 15 minutes. Reduce heat to 325 and bake 30 minutes more or until set. Garnish with pecan halves.

Dr. Franklin Martin's Less Sweet Potatoes

Peel sweet potatoes and eliminate any bad spots. Very carefully cut into ⅛-inch slices. Cover with water. Soak for 2 hours, stirring so all sides are covered with water. Discard water and rinse. Boil 20 minutes. Discard boiling water and mash. Add salt, milk, or margarine. Serve plain or mixed with beans, meats, or gravy, or mixed into soups. Can also be molded into patties and fried. (Best used fresh, so do not store for long periods.)

Peanuts:
A Southern Specialty
Contributed by Karen Smoke, Arcadia, Florida

Have you ever tried growing peanuts (*Arachis hypogaea*) in your garden? If not, you are missing out on a delicious taste treat as well as the chance to observe the flowering and fruiting cycle of this unique plant. Peanuts are often grown as a curiosity even by Northern gardeners, but a long, warm growing season is essential for successful peanut production.

Peanuts are a nutritious high protein food, matching or bettering, pound for pound, the protein value of eggs, dairy products, and most meats. They are an excellent crop for introducing children to the wonderful world of gardening. And, when plant residues are returned to the soil, peanuts add nitrogen and other nutrients. The roots of this legume affix nitrogen from the air. There are two types of peanuts, the erect bush and the runner or vining peanut. Plants have narrow to broad oval leaves. As they mature, showy yellow pea blossoms appear in the leaf axils, but these flowers do not become the peanut. They help to pollinate the inconspicuous flowers formed at the base of the plant on short stalks called pegs. The stalks elongate and push under the soil, where the ovaries enlarge and form the peanuts. This characteristic of flowering above ground and fruiting below ground is unique to the peanut.

Because the pegs must push into the soil, a soft, light-textured soil is preferable. Sandy soils with a pH of 5.8 to 6.2 are ideal. Peanuts require an ample supply of calcium, so apply dolomitic limestone or crushed oyster shells one season prior to sowing peanuts. Phosphorus and potassium are important for good root development. An application of wood ashes or 0-4-4 (N-P-K) fertilizer at planting time and again when the plants are 6 to 8 inches tall will boost yields. Because of their ability to gather nitrogen from the air, peanuts require no additional nitrogen fertilizer. Most varieties require a growing season of 110 to 150 days. Irrigate peanuts during the early growth stage but reduce moisture as the pods mature.

Peanuts must be planted in warm soil or the seed will rot. Seed may be sown either shelled or in the pod. Crack the pod very gently as you place it in the soil. Plant about 2 inches deep in rows spaced 1½ to 2 feet apart to allow space for cultivation. Seeds are spaced 6 inches apart in the row and later thinned to 12 to 18 inches. It is important to keep peanuts weed-free. Cultivate shallowly to avoid disturbing the roots, pulling the soil toward the stem as you hoe. Stop cultivating after the pods have begun to form and apply a mulch of hay or grass clippings. Mulch helps control weeds and keeps the soil surface loose, encouraging the pods to form near the surface, which makes harvesting easier.

Peanuts are harvested when most of the nuts are mature: at the first frost in temperate areas, or when the leaves begin to yellow and drop in the tropics. With a fork or potato digger, loosen the soil around the plant, and then lift the plant, root, "goobers" and all. Shake free of soil and stack with root uppermost in windrows to dry a few days if the weather is warm and sunny. If it is humid and wet when you harvest, remove the pods from the plants and dry on screens in a warm, airy place. Peanuts must be carefully cured and dried to 10-percent moisture content for storage, but you will probably want to enjoy your harvest fresh.

Sources for peanut seed are the Hastings Seed Company, in Georgia and Kilgore Seed Company in Florida. (See Appendix 3.)

Peanut recipes
Contributed by Betty Mackey

Roasted Peanuts
Place a single layer of peanuts (in shells) in a large shallow pan. Bake in a low oven (300-325 degrees) for 25 to 30 minutes. Stir several

times during baking. Sample after 20 minutes. The shells should be brittle and skins should slip off easily. Delicious served fresh and hot.

Boiled Peanuts

Green peanuts are used for boiled peanuts. Drop the peanuts into salted boiling water and simmer 30 to 40 minutes. Drain and serve hot.

Peanut Brittle

> 1 cup sugar
> ½ cup light corn syrup
> ½ cup raw peanuts
> ½ teaspoon vanilla
> 1 teaspoon baking soda

Cook the sugar, syrup, and water to the soft-ball stage (234-240 degrees). Add raw peanuts and cook until hard-crack stage (300-310 degrees). Syrup will be light brown. Remove from heat and quickly stir in the vanilla and soda. Pour onto oiled baking sheet, spreading as thin as possible. When cool, break into pieces. Store in an airtight tin.

Peanut Butter Pie

> ¼ cup peanut butter
> ⅔ cup dark corn syrup
> ¼ cup sugar
> 2 eggs, beaten
> 1 teaspoon vanilla
> 1 cup coarsely chopped peanuts
> 1 8-inch pastry shell
> Whipped cream topping

Blend the peanut butter and corn syrup in mixer bowl. Gradually add sugar, eggs, and vanilla. Place peanuts in bottom of pie shell; pour liquid mixture over nuts. Bake at 425 degrees for 10 minutes, then reduce heat to 350 and bake 25 minutes more. Cool and serve with whipped cream topping.

Cream of Peanut Soup

> 2 tablespoons flour
> 1 tablespoon butter
> 1 quart whole milk
> 1 cup peanut butter
> ½ teaspoon salt
> Finely chopped peanuts
> Whipped cream

Heat the milk. Make a roux of butter and flour. Stir in the milk gradually, blending well. Over very low heat, stir in the peanut butter until all is melted and smooth. Top each serving with a sprinkling of finely chopped peanuts and a spoonful of whipped cream.

Chuffa:
Nature's Earth Almond
Contributed by Margaret M. Cort, Largo, Florida

Chuffa (*Cyperus esculentus*) is a wholesome wild edible that can be grown year-round in a subtropical garden. In Florida's light and sandy soil, the tubers, which grow an inch or so below ground, can be easily removed.

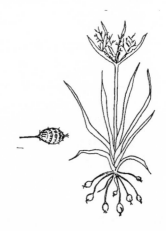

This "edible sedge" is readily identified by its grasslike leaves and stout, triangular reed stalk which grows to a height of 2 feet. A circle of leaves near the top resembles the bottom leaves in shape but the top leaves are smaller. The cluster of flowers is above the stem leaves and has 5 to 8 rays, each bearing many tiny, flat spikelets.

The tuber or nut grows underground and is the delight of the outdoor gourmet! When chewed, it yields a mouthful of sweet, milklike liquid. The flavor is distinctive yet similar to almond and/or coconut.

You can make hot or cold drinks, candy, puddings, and flour for cakes and cookies from chuffa. The Spanish *horchata de churas* liquid medicine was traditionally taken for digestive disorders and used by convalescents.

Over 3,000 years ago the Egyptians relished wild chuffa. Archeologists have found preserved chuffa in various pharaohs' tombs.

Chuffas are best planted in May or June and harvested in September, but in rainy years 2 crops can be produced.

Chuffa recipes

Chuffa Flour
Wash tubers and dry well. Set on cookie sheets in oven set at 300 degrees. Leave door open a crack to release moisture. When a cooled tuber breaks apart under a light hammer tap, it's ready for grinding. (It should take an hour or so.) Use food processor or grinder to make flour. Chuffa flour, mixed half and half with wheat flour, makes fine cookies, cakes, muffins, pancakes, and biscuits.

Chuffa Beverage
    ½ pound chuffa tubers
    4 cups water
    ¼ cup sugar

Soak tubers in cold water for 2 days. Drain. Put 4 cups water in blender and add sugar. Run blender on high, adding tubers gradually. When liquid is almost "homogenized," strain through a clean cloth. Serve either hot or cold.

For a coffeelike drink, roast chuffa tubers until dark brown. Pulverize in blender and brew just as you do regular coffee. Chuffa contains vitamin C and an important fat-splitting lipase enzyme.

Food-Bearing Vines

9

Growing Bunch Grapes in Warm Climates

Contributed by Bob Tillema, West Melbourne, Florida

Southern grape growing was once limited to muscadines, which grew as single large grapes. Northern and California grape varieties do not grow successfully in warm and humid areas. But now we have bunch grapes especially for the South.

Growing your own grapes is a very satisfying and rewarding experience. Most adaptable varieties produce many pounds of grapes in 2 to 3 years. Most are delicious eaten fresh and some are extra good for wine and jelly.

Select a site with well-drained soil. Most light, sandy soils with pH of 5-8 are suitable. It is best to run rows from north to south for adequate sunshine on both sides of vines. Space vines 8 to 9 feet apart or closer if more dense foliage is desired. Leave 10 to 12 feet between rows.

Trellis grapes to keep shoots off the ground. Set 8-foot-long treated 4x4s 2 feet deep. Staple galvanized wire or plastic-coated nylon rope horizontally at the 3½-foot level and on top of the posts at 6 feet. Select a strong shoot and tie it to the stake for a permanent trunk. Remove other shoots, but leave one lateral shoot to grow each way on each wire. Cut off vine when it reaches top wire. Two or more laterals will form, which will be left and right arms on top wire.

Pruning is done during the dormant season. We recommend early January pruning in the tropics. Remove shoot growth less than ⅓-inch thick each year. Allow only sufficient buds to produce a few bunches during the first year, 5 pounds the second year, and 10-15 pounds after the third year. Calculate pounds by multiplying the number of buds by 2 bunches per shoot by the approximate weight of each bunch. Stover bunches weigh only about 4 ounces while Roucaneuf bunches weigh almost 1 pound. So consider your grape variety when pruning. Vines should be pruned 70 to 80 percent each year.

Fertilize grape vines 3 times each year. Make first application when spring growth appears. Use a 20-20-20 with minor elements. In May and August, use 3-2-2 ration with added elements. During the first year, a monthly application of water soluble product (20-20-20) will get your vines off to a great start. Do not fertilize after August.

Mulch to keep weeds down and to maintain better soil conditions. Irrigate during dry periods, weekly, and more often during the first year.

For maximum production use a spray program. But organic gardeners know that very well-nourished grape vines will avoid diseases and insect problems.

The Isabella grape has been widely grown in Hawaii since 1800, mostly by the Portuguese there. The Thompson Seedless 52-838 variety is the most extensively planted grape in California. It requires plenty of heat and is the best variety for hot, desert areas. It is vigorous and productive. Flame, Muscat of Alexandria, and Lady Finger are good seedless grapes that grow well in California-type climates. The Amazon Tree grape produces well in very hot, moist climates.

Grape recipes

Grape Juice

My favorite way to extract juice from grapes is by using a steam juicer. Developed in Europe, this device is a giant double boiler. Grapes are cooked over hot water. Pure juice runs out of a little plastic tube into bottles you have rinsed with hot water. Then capped juice bottles can be kept safely at room temperature. (Look for steam juicers at specialty cookware shops.)

Also make juice by washing fruit. Put in large kettle with water to cover. Mash fruit and cook until fruit is very soft. Pour into cloth jelly bag and hang for juice to drip out. Then boil for 3 minutes, pour into sterilized jars and seal at once.

Always remove grapes from stems before cooking or juicing. Grape wood gives you a bitter flavor.

Grape Butter

    3 cups grape pulp (left over from juice making, with
            seeds removed)
    3 cups sugar

Cook slowly until thick. Stir often to prevent burning. Pour into hot sterilized jars and seal.

Grape Jelly

    3 cups juice
    2 to 3 cups sugar

Place juice in 4-quart kettle. Boil rapidly for 5 minutes. Add sugar and bring to jelly stage (224 degrees). Skim off foam and pour into hot sterilized jars. Makes 10 4-ounce glasses of jelly. These are baby food jars which make lovely little gifts.

Sugar-Free Grape Jelly

> 2 cups grape juice
> 4 teaspoons unflavored gelatin
> Noncaloric sweetener to equal ½ cup

Soften gelatin in ½ cup grape juice. Bring remaining juice to boil. Add softened gelatin. Stir to dissolve. Add sweetener and bring to boil. Remove from heat. Pour into container and store covered in refrigerator. (From *Rare Fruit & Vegetable Council of Broward County Cookbook*, 1982.)

Grape Dumplings

> 3 to 4 cups washed and stemmed grapes
> 2 cups flour
> 1½ teaspoons baking powder
> ¾ teaspoon salt (optional)
> 1 tablespoon soft butter
> ¾ cup milk

Cover grapes with water and simmer until tender. Rub through sieve to remove seeds. Sweeten to taste with sugar or honey. Mix flour, baking powder, and salt. Add butter and work mixture with pastry blender. Add milk. Turn out on floured board and roll ½-inch thick Cut into small squares. Use a heavy pan to bring grape juice and seeded pulp to boil. Drop in dumplings one at a time. Cover and cook about 15 minutes until dumplings are done. (This recipe was a favorite of Native Americans who used wild grapes to make it.)

Spiced Grapes

> 8 pounds grapes
> 4 pounds sugar
> 3 cups vinegar
> 4 sticks cinnamon
> 1 ounce whole cloves
> 2 blades mace

Slip skins off grapes. Cook pulp with spices tied in cheesecloth until grapes are soft. Mash through sieve to remove seeds. Add skins. Bring to boil. Add sugar and spice bag. Cook until thick. Put into jelly glasses and seal. See Chapter 12 for wine recipe.

The Kiwifruit:
An Astounding Producer

Parts contributed by Dr. Jurgen Reckin, Experimental Garden, Finowfurt, Germany

The kiwifruit, also called Chinese gooseberry, strawberry peach, and yang-tao, is native to China and is botanically known as Actinidia chinensis. It has become so popular during the last decade that more people in the subtropics and in warmer parts of the temperate zone are now trying to grow them. However, the vines may not fruit in areas that are too hot and moist.

The climbing vines of this shrub bear heavy loads of oval, often nearly cylindrical berries about the size of an egg. The fruit is covered with soft, reddish-brown hair. Some varieties bear fruit the second year after planting, but others grow 5 to 8 years before fruiting. One vine may produce over 300 pounds of fruit! The plant is dioecious and 1 male shrub per 10 females is enough for good pollination. The creamy-white, pleasant smelling flowers are borne in clusters.

A cross section of the fruit shows an unusual, colorful picture: the juicy pulp is medium green with lighter radial stripes and contains many tiny black seeds. The innermost part is creamy white. The taste, difficult to describe, is somewhat like a mixture of gooseberries, strawberries, and peaches.

The fruit is mostly eaten raw or in fruit salads. Its vitamin C content ranks among the highest in fruits; depending on varieties and growing conditions it varies between 60 and 600 mg of vitamin C per 100 grams of peeled fruit. Recent studies have found that one large kiwi has 89 mg of vitamin C, which is more than an average orange. With 322 mg of potassium, it has as much of this important mineral as a banana. It also contains as much fiber and iron as an average apple. The fruit stores well and can easily be shipped long distances. New Zealand supplies fruits during our summers and California markets them during the winter.

New Zealand breeders continue creating new varieties. The Hayward bears large oval fruit that is slightly flattened laterally, pale greenish-brown, and densely covered with fine hairs. It is superior in flavor and keeping-quality. Other good varieties are Abbott, Bruno, Monty, and Gracie. Grafting is preferred for propagation of named varieties, but there are also reports of successful rooting of hardwood and softwood cuttings.

Kiwifruit requires some winter chilling in order to fruit. In spring when new growth occurs it is frost-sensitive. Deep, free-draining soils are important for successful culture.

Remember that kiwifruit need a rich, well-drained soil with a pH from 5.5 to 6.8, average watering, annual winter pruning, light shade to full sun, and a very strong support. A single well-established plant, female with male grafted on it, can grow to 30 by 30 feet and produce 300 pounds of fruit in one season. However, as far as we know, kiwis have not produced fruit in south and central Florida. We tried growing the Vincent, which does very well in San Diego, but we could not get it to bloom in Melbourne. But growers in northwest Florida have fruited kiwis. They are now trying the new California variety. There are also several kiwis with tiny fruit that produce well in the North. We hope kiwifruit fruit will be developed so we have a variety for every climate.

You can plant the kiwifruit seed from the fruit you buy in grocery stores, but they should be kept in your refrigerator at 40 degrees for about two weeks before planting. They should do well as seedlings in

70 to 80 degree weather. They need full sun and should be kept well watered and fed.

Some California growers have grossed $12,000 per acre on their kiwifruit. Kiwifruit are growing successfuliy in Dawes, Alabama, and the fruit is doing well in South Carolina, which has set up a state marketing board for the crop. The biggest problem with producing kiwi vines is making arbors strong enough to hold up all the fruit!

Kiwifruit recipes

In the summer of 1980 almost every major U.S. women's magazine had a cover featuring a luscious dessert made with kiwifruit. While I prefer kiwifruit fresh without added sugar, here's a good torte recipe:

Glazed Kiwifruit Torte
> 4 eggs, separated
> 1 cup mild-flavored honey
> ½ teaspoon cinnamon
> 1 teaspoon vanilla
> 1 cup fine soda cracker crumbs
> 1 cup finely chopped walnuts

In a large mixing bowl beat egg yolks until thick. Add honey in a fine stream while you continue beating. Fold in cinnamon, vanilla, cracker crumbs, and nuts. Beat egg whites until stiff and fold them into egg yolk mixture.

Cut a round of wax paper to fit bottom of an ungreased 9-inch round pan. Pour batter into pan and bake at 325 degrees for about 20 minutes or until center is done. Cool. Carefully put torte on a pretty glass platter.

Make topping with:

> 2 cups chilled evaporated milk
> 3½-ounce package instant vanilla pudding

¼ cup sherry wine or fruit juice

Using small mixer bowl, whip milk until it just holds shape. Turn off beater and sprinkle pudding mix over cream. Continue beating until mixture holds firm peaks. Gently fold in sherry. Spread over torte. Peel 2 kiwifruit and slice. Peel 1 banana and cut in rounds. Dip fruit slices in honey-sweetened lemon juice. Arrange over cream topping. Makes 8 to 10 servings.

The Passion Fruit:
For Juice You Will Really Love

Contributed by Arnold Pechstein, Sebastian, Florida

There are over 600 species of passion flowers in the genus Passiflora. Some of these are shrubs and small trees, but most are vines native to the Americas.

The flowers vary in size and shape, and colors range from pure white and brilliant red to delicate shades of lavender and blue. Botanists rate some as the most beautiful blooms in the world! The fruit of *Passiflora suberosa* is the size of a blueberry and practically tasteless, while that of *P. quadrangularis* is 12 inches long, shaped like a melon, and has a very delicious flavor.

Of the genus *Passiflora* there are probably 40 or 50 species bearing fruit considered worthy of propagation. Rare Fruit Council International of Miami, Florida, lists 11 species of *Passiflora* in their yearbook. Because of Miami's low elevation, many of the highland varieties will not survive there, but should have a better chance in California and other areas.

Nutritionally the passion fruit rates well and contains carbohydrates, calcium, phosphorus, and some iron. It has become a commercial fruit in many parts of the world, used for making soft drinks and wine and simply sold as a fresh fruit.

Passion fruit could become a commercial fruit in more areas if certain problems were overcome. Pollination to get fruit set is one problem. Studies in Hawaii with *P. edulis* and *P. flavacarpa* indicate certain seedlings are not self-compatible and will only set fruit when pollinated with pollen of other seedlings.

Since passion vines grow quite readily from cuttings, we could grow vines we know are self-compatible. Studies show honeybees are only 25 percent effective in pollination of yellow passion fruit, with carpenter bees about 50 percent effective. Hand pollination has produced 100 percent fruit-set on certain seedlings. This is probably why commercial growing of passion fruit has been limited to areas where labor is cheap. Hybridization and selection could possibly give us more varieties more readily pollinated by honeybees.

Purple passion fruit (*P. edulis*) is affected by crown rot, which kills the plant. However, this can be overcome by grafting onto the yellow passion fruit, which is resistant to crown rot.

Caterpillars are often a problem on young plants. Butterflies of the *Colaenis julia, Colaenis delia,* gulf fritillary, and zebra varieties lay eggs on terminal growth of vines. But I've never had caterpillars on full-grown vines. I wonder if the little red ants eat these butterfly eggs.

Passion vine varieties flower and fruit at different times. In Florida most fruit during summer months. But Scarlet Passion Florier (*P. coinea*) begins flowering in fall and fruits until spring or midsummer. Purple passion bears fruit for about a month in spring, while yellow passion bears from July until December.

I have a cross between the yellow and purple which flowers and fruits, year-round, being self-compatible from October to June and requiring the pollen from another yellow seedling from June to October. Unfortunately, this one suffers from crown rot and needs to be grafted.

Growing passion vines can be enjoyable, but it would be a lot more fun if we had all 40 or 50 species that bear a good edible fruit here in the United States. So when you make a trip to the tropics and find a passion fruit in the market, bring back the seeds and plant them. (Make sure you have a seed import permit. See Appendix 3.) It may turn out to be a new species or it could be an old one with characteristics we want for good growth. Remember, you never know what you will get when you plant a seed. This makes growing things very exciting!

You may wonder how one could use the passion fruit, as its guava-like shell is much too hard to eat. But it's easy. Just cut fruit in half, scoop out seedy pulp, and press through a wire strainer. The pulpy juice can be mixed with orange, pineapple, grapefruit, and other juices for a very fine drink. With added pectin it can be made into jelly. It can also be made into chiffon-type pies, fruit sherbets, and ice cream. Once you start growing your own passion fruit, you will certainly be moved to create your own recipes. Shells go in your compost pile.

Chayote:
The Vegetable Pear

There is so much to say about the chayote. Botanically known as *Sechium edule,* this member of the squash family originated in the West Indies and is grown outdoors as far north as Oklahoma. It was called *chayotli* by the Aztecs and is known as *chocho* in South America and *miriton* and *christophine* in Creole cookery.

The chartreuse, pear-shaped fruit, which often becomes quite wrinkled, comes in several varieties differing slightly in shape and color (from white to dark green). The leaves can be heart-shaped, angled, or lobed.

The cucumberlike vine is a perennial and in colder climates is commonly protected by an inverted bushel basket. Most experts say the chayote vine must have support, but a neighbor has grown fine chayotes on flat pine-needle beds.

The large tuberous roots can grow to 20 pounds and are often eaten after a plant has grown for several years. The young shoots and leaves are cooked like spinach and even contain the same oxalic acid. Folklore says ancient royalty were treated with the toasted large single seed, which has a slight almond flavor. Personally, we think the seeds are delicious too!

The whole fruit is usually planted horizontally and lightly covered with soil. We often wait until it has a sprout several inches long before planting. We've had best results planting chayotes in soil enriched with mulch, sludge, and well rotted-manure with some added gravel or lime. The chayote seems to thrive in the Miami area, where the oolite layer gives it plenty of natural calcium.

The chayote flowers and sets fruit only during the short days of fall. Sometimes only male or female flowers occur on a single vine, so it's best to plant several chayotes at the same time.

A neighbor very carefully cut off both cheeks of a mature chayote fruit, ate them, and planted only the sprouted seed—with very good results.

Chayote recipes

Young tender chayotes may be sliced and served raw in salads. You can use them in souffles or cook them au gratin—but watch the skins as they get older. (The first time I cooked them in a casserole for guests, we had to spit out the skins!)

Baked Chayotes
Use ½ chayote per person. Cut chayotes in half lengthwise and put cut side down on buttered or oiled baking sheet. Bake at 350 degrees until they test done with a knife. Serve hot with salt and pepper to taste.

Stuffed Chayote

Contributed by Ollie Porto, Sharpes, Florida

Wash chayote. Cut in half lengthwise. Scoop out seed and membranes. Stuff with dressing of your choice, enriched with chopped cooked pork, chicken, or sausage. Bake at 350 degrees until tender. Serve one half per person, with a green salad.

Chayotes Stuffed With Shrimp

Wash 6 large chayotes, cook in lightly salted boiling water until just tender, and drain. Halve lengthwise. Carefully scoop out pulp, chop finely, and drain off excess liquid. In a large pan saute 1 large onion, chopped, in 1 tablespoon butter until translucent. Add chayote pulp, 2 cups bread crumbs, and oregano, salt and pepper to taste. Simmer mixture over low heat, stirring often, for about 10 minutes. Add cooked shrimp, shelled and minced. Heat thoroughly. Fill chayote shells with shrimp mixture. Sprinkle with bread crumbs and dot with butter. Put in large shallow baking pan. Add ½-inch hot water. Bake at 350 degrees until golden brown.

Maria's Dessert Chayotes

Cook 6 large chayotes in lightly salted boiling water until just tender and drain. Halve lengthwise and carefully scoop out pulp, leaving a thin shell. Chop pulp finely and drain off any excess liquid. Combine pulp with 3 cups milk, 3 well-beaten egg yolks, 4 tablespoons brown sugar and ¼ teaspoon vanilla. Fill chayote shells with mixture. Put in large shallow baking dish and add ½ inch hot water. Bake at 300 degrees until filling is set and top lightly browned. Serve hot as a dessert.

The Famous Calabaza

Probably originating in Mexico or Peru, the calabaza or Cuban squash (*Curcubita moschata*) is one of the very best food producers for hot summer weather. One seed will produce many pounds of food!

About 30 years ago a former neighbor, Wayne Taylor, started growing them in the area. He shared the seeds with many of us—and with the help of our neighbors we've kept it going ever since.

We sent seeds to Jack's father in El Cajon, California, and he grew a gigantic crop of insect-free calabaza. He furnished a Thanksgiving dinner vegetable and dessert for his church, which had over 100 members.

Major seed companies have not been interested in propagation, probably because they thought calabaza was limited to very warm climates. But we received a report from one of our readers who has grown a large crop in Faribault, Minnesota.

The plant has squashlike leaves with specks of white, and the young green-and-white striped fruit becomes buff-colored when fully ripe. The fruit can be pumpkin, pear, or watermelon-shaped. The blossoms can be eaten dipped in batter and fried. The young fruit is excellent steamed, and the mature fruit, when baked, tastes more like sweet potatoes than squash.

We get best results when planting the squash in full or partial sun where it has plenty of room to spread. Sometimes the vines grow to 60 feet. Be sure to prepare the planting area ahead of time by digging in plenty of manure, compost, and fertilizers. Mulch heavily around the plants with pine needles, grass clippings, or straw. Insects have never been a problem.

Our best Florida crop resulted when seeds were planted in March or in mid-August. We distributed more than 100,000 seeds through my newspaper columns, radio and TV, and organic gardening and national gardening magazines. They were sent all over the world. Reports came back telling of success. One woman in Oklahoma wrote, "From the three seeds you sent, we had a pickup-truck load of calabaza!" A resident of North Carolina reported, "The calabaza were insect-free and we had no problem with squash bugs."

The biggest asset is that they grow well during long, hot summers, make a fine ground cover to keep out weeds, and also produce a nourishing food crop. Mature calabaza will sometimes keep for 6 months at room temperature in the subtropics. The shells are very hard, and sometimes we use a small saw to cut ours.

In 1988 a new seed company, Southern Seeds, in Melbourne, Florida, decided to market our calabaza and named it the Marian Van Atta.

Calabaza recipes

Mary Wood's Calabaza Casserole

        3 tablespoons butter or oleo
        1 cup milk
        1 cup dry bread crumbs
        2 cups cooked calabaza, mashed
        1 tablespoon grated onion
        1 tablespoon canned pimento
        1 teaspoon salt (optional)
        ⅛ teaspoon pepper
        2 eggs
        ½ cup buttered bread crumbs (or wheat germ)

Preheat oven to 350 degrees. Melt butter in hot milk and pour over bread cumbs. Mix well. Add squash, onion, pimento, and seasonings. Beat eggs and add to mixture. Pour into oiled 2-quart casserole and cover with buttered crumbs. Bake 20 to 30 minutes. Serves 6.

Calabaza Pie

Use your favorite pumpkin pie recipe and simply substitute the cooked calabaza for the pumpkin.

Rancho Luna Calabaza

One 5- to 6-pound calabaza. Cut out center to 1 inch of outer skin. Dice center pieces. (Make sure you save enough seeds to plant and share with your neighbors.)

Mix diced calabaza with:

> 1 tablespoon light rum
> 1 tablespoon grated coconut
> ½ cup raw rice
> ¼ cup diced green pepper
> ½ pound small cleaned shrimp
> Salt and pepper to taste

Return mixture to calabaza shell. Put in large oiled baking dish. Bake covered 1½ to 2 hours. Before last 20 minutes of baking, remove cover and place about a dozen unshelled shrimp on top. Just before serving, add several dashes of rum and cover until ready to bring to table. Then uncover and garnish top with lime wedges.

Calabaza Omelet

> ⅔ cups cooked calabaza, mashed
> 3 eggs
> 1 green pepper, chopped
> 1 onion, chopped

Saute pepper and onion in peanut oil until soft. Beat eggs and combine with pepper, onion, and calabaza. Cook with low heat in a heavy skillet. Cover last 5 minutes or until done on top.

Kiwano:
A Controversial Melon

In 1985 a strange-looking fruit appeared on the U.S. market. Botanically known as *Cucumis metuliferus,* it was shipped from New Zealand. Apparently growers of the kiwi, which had become very popular, were looking for another successful introduction. The brightly circled, orange fruit looked very odd and attracted a great deal of attention. A TV personality even ate a whole fruit on a national a show. It is now available in markets all over the United States.

Also known as the African horned melon, the kiwano was used as a food only in times of famine in Africa. While the flesh is traditionally green, much like the cucumber, it does not have much flavor. Several companies developed recipes for it, but because the seed was too much trouble to separate from the small amount of juice or jelly-like material, many felt the fruit was not worth trying to eat. However, it makes a striking addition to an arrangement of fruit and flowers—especially autumn displays. When the shell is scooped out, it can also be used as a container for salads and puddings.

Then an environmentalist began investigating. It seems the spines were sometimes ground off before shipping. It was also considered a serious weed in Australia. Several folks in Florida tried growing the fruit and only a few found it thrived. Others were barely able to get it to fruit.

Reports of plant breeders trying to improve the flavor and remove the spines are encouraging. It will give this unusual fruit more of a chance to win the hearts of exotic food lovers sometime in the future.

Because there is always a possibility of the kiwano escaping and going wild, we suggest that it not be planted in its present form. It could be worse than sandspurs in Florida lawns!

Useful Plants from Seeds and Cuttings

10

Roselle:
The Florida Cranberry

Let's start a revival: let's all grow roselle. If you live where you have 8 or 9 months free of frost, you can! Also known as Florida cranberry, Jamaica sorrel, and botanically as *Hibiscus sabdariffa,* this plant was popular in the subtropics during the thirties. Now food prices are so high we should use it again. It has many benefits. Roselle can be used to make jelly, juice, fruit pies, and tarts. You can even use the tender green leaves and stems as you do spinach.

One of the reasons roselle has lost favor is that this plant is an annual and needs to be replanted every year, but then so do tomatoes! It grows into a fine-looking hibiscuslike shrub 3 to 5 feet high and can be used as background planting in home landscaping besides being placed in rows. It makes quick shade for the summer and in winter often freezes back, clearing the area for solar heat.

The green leaves begin as compound but as they mature change to palm-shaped with 5 parts. The stems are stiff and reddish-colored. Flowers resemble the hibiscus, a close relative, and in the most common variety are pale-yellow with a dark-red center and golden pollen. Flowers are open only a day. Most confusing is that you use the calyxes or little cups that hold the petals. There is no other berry or drupe. Soil should be prepared ahead of time with mulch and well-rotted manures. Seeds should be planted in spring, 3 or 4 seeds to a hill. After 2 or 3 ordinary leaves develop, plants should be thinned. Plenty of heavy mulch keeps weeding down. Additional feeding is usually not needed but plants should be kept well watered. Roselle will not bloom until daylight hours are short, so crops ripen in fall and winter. The *Philippine Agricultural Review,* 1920, lists several varieties of roselle. They include: Rice—with plants less tall and more spreading; Victor—taller with more slender, tapered calyxes and earlier fruiting; Archer—robust with green-and-whitish leaves and stems and greenish-white calyxes producing amber-colored products; Temprana—an early fruiter; and Altisima—robust with larger but inedible fruit and used for the long silky fiber extracted from the stems.

Roselle is subject to root-knot nematode and should not be planted in land infested with this pest. Caterpillars are sometimes a problem; they can be controlled with Dipel. We have found that Roselle stops producing edible calyxes when planted in the same area for several seasons. It may be a problem to change planting areas every year, but this unique fruit is worth the extra effort! Calyxes mature very rapidly and are ready to harvest about 15 days after blooming. Pick them when they snap off easily. Plants will bloom and produce more when kept well picked.

Roselle juice looks much like cranberry juice and has much the same food value, except roselle doesn't contain the benzoic acid that gives the bitter taste to cranberries. Both contain malic acid and should not be cooked in thin metal containers.

According to a University of Miami scientist: "Roselle juice is antibiotic, diuretic, relieves fever and coughs, stimulates intestinal activity, and is effective in lowering blood pressure."

Wine has been made commercially from roselle herbage in the Philippines. You can follow the basic wine recipe in Chapter 12.

Dried roselle is available in produce departments in southern California and Mexico and also as "dried hibiscus" in many health food stores.

Roselle recipes

Roselle usually begins to bear in November and then lasts for several months. Make sure to keep your calyxes picked when they snap off easily with your fingers. If you have to use a knife to remove them, you've let them go too long. They can still be used for juice but are better when young.

For all but the roselle jelly and beverages, the seedpods need to be removed (left in, they are too hard on teeth). Just cut off stem, slit side of calyx, and remove seedpod with your thumb. (Get help or you may get a sore thumb!)

Roselle Juice

Pick and wash whole fruit. No need to remove seedpod for juice. Put in large kettle with water to cover. Boil about 10 minutes until tender. Strain through a cloth jelly bag.

Roselle Jelly

    3 cups roselle juice (prepared by above method)
    2 ¼ cups sugar

Bring juice to a boil. Add sugar and continue boiling to 222 degrees or until it passes jelly test. Pour into sterilized jars and seal.

Roselle Punch

    4 cups roselle juice
    ½ cup sugar or honey

1 cup orange juice

Juice of one lemon or lime

Dissolve sugar or honey in juices and serve over ice.

Hot Roselle Punch

6 cups roselle juice

1 cup sugar (less if desired)

1 cup orange juice

1 lemon (rind and juice)

1 tablespoon each whole cloves, allspice, and
stick cinnamon

Combine roselle juice and sugar. Add spices and lemon rind. Simmer for 10 minutes. Remove from heat and strain. Add orange and lemon juice. Heat again and serve. This punch makes a festive holiday drink.

Roselle Sauce

3 cups roselle calyxes, seeded

1 cup water

Sugar or honey to taste

Add water to calyces and cook until soft. Add sugar (about ½ cup for each cup roselle) and cook until dissolved. Your favorite spices may be added. Or use part vinegar and water. Great served either hot or cold with poultry and meats.

Roselle Turkey Stuffing

2 cups roselle calyxes, seeded

1 cup chopped pecans

½ cup sugar or honey

1 cup each chopped celery and onions

¾ cup milk

2 quarts soft bread crumbs
(you can use a stuffing mix)

½ cup butter or margarine
1 teaspoon salt (optional)
¼ teaspoon each: sage, thyme, and black pepper

Combine roselle, pecans, and sweetening. Set aside. Heat butter in a skillet. Add onions and celery. Cook over medium heat, stirring to prevent burning. Add salt and herbs. Mix bread crumbs, roselle mixture, and onion and celery mixture together. Add milk. Spoon stuffing into neck and body cavities of turkey. Extra stuffing can be put in oiled baking dish and baked for last hour of roasting time.

Roselle Tarts

4 cups roselle calyxes, seeded
1½ cups sugar
½ lemon rind (grated)
1 tablespoon flour
½ teaspoon cinnamon
⅓ cup water
2 tablespoons butter
½ cup chopped walnut or pecans

Mix sugar, flour, and cinnamon with water. Heat until sugar is dissolved. Add roselle and cook until tender. Add lemon rind and butter. Pour into pastry-lined tart pans. Top with lattice strips. Bake in 400 degree oven for 30 minutes.

The Poha Berry:
Reach for the Gold
Contributed by Gloria Glenn, Cocoa, Florida

Physalis peruviana, or the poha berry, is a native of Brazil but has become a favorite fruit of Hawaii. It is a member of the Solanaceae family which has many subspecies. Also called cape gooseberry, ground cherry, husk cherry, and goldenberry, poha's fame and many uses have spread from the tropics. Folks now grow it during the summer in cooler climes. The fruit resembles a ⅜-inch x 1-inch greenish-yellow tomato and produces lantern-shaped papery calyxes to keep insects out. It tastes a little like pineapple. John Riley, a California solana expert, says it dries into a wonderfully delicious raisin. The plant grows to about 2 feet in height with spreading branches and heart-shaped, shallow-toothed leaves.

It is easy to cultivate. To extract seed, squeeze berry into a glass of water. Stir and allow seed to settle. Pour off water and strain dregs through a paper towel to catch the seeds. John says you can sow seeds on a vacant lot for a carefree crop or grow them in your garden exactly as you would tomatoes. When fully ripe the pohas may be eaten raw. In Hawaii it is made into strawberry-type shortcakes. Because the fruit is low in pectin, it is not good for jelly but can be made into a fine jam.

"Over 2,000 species of solana are presently in existence and only a small number have been exploited," John says. "I'm quite fearful we're going to destroy their habitat before they're all discovered."

Poha recipes

Poha can be eaten fresh and in fruit salad, or cooked in dishes like these:

Poha Jam
      4 quarts husked pohas
      1 cup sugar to each cup (5 to 6 cups) cooked pohas

Husk, wash, and cook pohas slowly for 30 minutes. Stir frequently until there is sufficient liquid to prevent fruit from scorching. Let stand overnight. Measure poha pulp and juice and add an equal quantity of sugar. Cook slowly 30 minutes to an hour, stirring mixture frequently until juice thickens when cooled. Pour into hot sterilized jars and seal. (The quantity of juice varies according to climate. When water content is unusually high, pour off some of the juice before adding the sugar if more fruit than jelly is desired.)

Apple-Poha Berry Pie

    4 tart apples
    ¾ cup sugar
    Cinnamon or nutmeg
    Halved poha berries
    Grated lemon rind

Use your favorite 2-crust pastry recipe. Spread thinly rolled pie crust on bottom of pan. Arrange half the apples on bottom crust, then enough poha berries to fill in the spaces. Sprinkle with blended sugar and flavoring. Top with remaining apples and berries. If apples are dry, add 1 or 2 tablespoons water. Flavor as before. Wet edges of crust, cover with top crust, and cut two or three gashes in it to allow steam to escape. Bake in 450-degree oven for 10 minutes, then reduce heat to 350 and bake about 30 minutes longer. Makes one 9-inch pie.

Pineapple:
Growing Your Own Mini Plantation

Parts contributed by Myrtle Pell Edmon, Bellflower, California

Could you spare a piece of land about 6 by 15 feet to have your own fresh pineapple? With proper cultivation you could have fruit in less than 18 months.

If possible, select an area of virgin soil under light shade. Prepare soil by rototilling. Enrich with manures, commercial fertilizer, bonemeal, and a handful of iron sulfate. Also add seaweed— used by early Florida pineapple planters to grow superb crops. Experts say it is best to use either ratoons, suckers, or slips—not crowns—for planting. However our neighbor, Dr. Charles Griffith, started his fine pineapple plantation with crowns from store-bought fruit. Place young plants 15 inches apart in rows 20 inches apart. Don't set in too deeply, just enough to keep the plants from falling over. Choose a good mulch such as building paper, black plastic, pine needles, or sawdust. Plants should be watered frequently but need good drainage. Feed four times yearly until plants bear fruit; then feed twice a year, keeping material on ground and off plant. Use any organic materials including coffee grounds, worm castings, and compost. Also spray plants monthly with a seaweed nutritional supplement. Protect your pineapple bed with a canvas tarp when the temperature goes below 32 degrees. The bed may last 6 to 8 years if you care for it well and weed out suckers and old plants.

If your pineapple plants don't bloom by 18 months, you can try "gassing." Put water in bud. Drop 10 grains calcium carbide (available at welder supply stores) into bud. This should cause plants 18-20 months old to bloom. Fruit will mature during next 5 to 7 months.

The pineapple originated in South America. Even now one species, *Ananas microstachys,* grows wild in Paraguay. Natives call it the "Na-na," meaning "fragrance." Varieties include Smooth Cayenne, Sugar Loaf, Abaco, Red Spanish, Queen, and others.

The Pineapple's Secret

Pineapple fruit may resemble a hand grenade, but instead of containing high explosives that destroy, the pineapple contains a delicious substance that aids digestion. Like any fruit, if left to ripen on the plant it will have a sweeter, superior flavor. Pineapple (*Ananas comosus*) is a tropical perennial and looks like the agave or yucca. It grows from 2 to 4 feet high. Pineapple is easy to grow but requires patience. Like the asparagus, it takes about two years to produce fruit. There is no need to buy an expensive nursery plant. Simply start with fresh fruit found in the market. Select a pineapple with a green center in the leafy top (crown). Twist off crown. Start at bottom of crown and pluck leaves one by one about an inch up the stalk. The stalk will be a rich cream color. Hang upside down in a dry, cool place for about 5 days to allow cut end to harden to prevent rot.

Then plant in an 8-inch porous clay pot with good drainage. Cover hole with a piece of broken pottery. Add a small amount of coarse gravel. Plant crown about 1 inch deep in equal-parts mixture of light garden and potting soil. Tamp down firmly, leaving green leaves exposed with no soil on leaves.

Pineapples like sun but not frost. Cover plant or take inside during cold weather. Water soil once a week. Every 2 or 3 months feed plant directly on soil—never on plant. If pests appear, treat as any other garden plant.

Be patient. It takes crown at least 26 months or longer to produce fruit. As the plant grows, transplant it into larger pots. When about 16 months old, a deep pink bud begins to form in center of leaves. In 2 more months, a bright red cone will appear.

If plant is 20 months old and hasn't bloomed, place a ripe red apple next to plant, tie a plastic bag securely over entire pot, and place it in the shade. Remove bag in 3 to 4 days and return plant to usual sunny location. In about 2 months the red cone will appear, and 2 weeks later bright blue flowers will open row by row. Then fruit begins to develop.

When fully developed, the green pineapple will turn a rich gold halfway up. Now it is ready to be picked, eaten, and enjoyed. I'm sure you will say, "It was worth the long wait."

Pineapple recipes

Pineapple Juice
If you have a fruit and vegetable juice extractor, run peeled, ripe pineapple through. If not, peel and core fruit. Finely chop and squeeze through a strong cloth bag.

Pineapple Shrimp Cocktail

    1 cup cooked shrimp
    2 cups diced fresh pineapple
    Lime juice

Combine all and serve in sherbet or cocktail glasses.

Pineapple Cabbage Slaw

    3½ cups shredded cabbage
    2 cups fresh pineapple, shredded
    1 green pepper, chopped
    ½ cup salad dressing

Combine all and serve chilled.

Hawaiian Dessert

    2 cups pineapple chunks
    1 cup ripe papaya, cubed
    1 banana, sliced
    1 cup grated coconut

Combine all. Chill and serve.

Pineapple-Sweet Potato Fritters

    2½ cups sweet potatoes, mashed
    2 eggs
    4 tablespoons melted butter
    1 cup crushed pineapple
    1 cup flour
    1 teaspoon baking powder
    ⅛ teaspoon nutmeg

Mix together potatoes, egg, butter, and pineapple. Add flour sifted with baking powder and nutmeg. Mix well. Drop by spoonful into hot, deep fat. Set fryer at 375 degrees. Fry until golden brown.

Chinese Barbequed Pork

¼ cup soy sauce

2 tablespoons salad oil

2 garlic cloves, mashed

1 small chili pepper, crushed

¼ teaspoon anise seed

⅛ teaspoon each of cinnamon and cloves

1 pound pork, cut into cubes

1 fresh pineapple

1 green pepper

Combine first 7 ingredients in bowl. Add pork to marinate. Put pork on skewers with pineapple and green peppers. Grill 7-10 minutes. Serve hot.

Aloe:
Ancient Healer Used Today

Contributed by Norman Lund, Melbourne, Florida

Aloe is the oldest known and most used medicinal plant in the world. It was used by ancient Egyptians, Babylonians, Syrians, and Aztecs. The conquistadors brought it to the New World.

Unfortunately, it does not withstand cold and is grown outside only in tropical and subtropical regions. However, it can be grown in a pot as an indoor ornamental. The succulent leaves can be carefully cut off from the bottom of the plant. Then the oozing gel can be used as a healing aid for burns, bruises, skin abrasions, acne, sunburn, insect bites, infections, and as poultices on sore muscles.

Aloe is a succulent member of the lily family. It can vary in height with some types growing less than a foot and others growing to tree size. Leaves are thick and tapering and can have smooth or spiny edges.

All grow in rosette form. Popular houseplant varieties include *Aloe arborescens, A. aristate, A. saponaria,* and *A. variegata.*

But few understand what in the aloe gel promotes healing. Aloe gel is 99 percent water. The remaining 1 percent contains more than 50 ingredients including aloin, proteins, amino acids, saponins, glucose amines, vitamins, enzymes, quinones, and minerals (calcium, magnesium, barium, zinc, and bismuth). Most important is a naturally occurring polysaccharide known as polyurinide. In combination with calcium, it works as a pain reliever and healing agent. The polyurinide rapidly detoxifies and relieves pain. After detoxification rapid healing begins.

When a leaf is cut and put in the refrigerator, the cut end will seal and the gel inside will retain poteny for several days. But once removed from the leaf, the gel deteriorates rapidly due to enzyme action. The gel may be preserved by putting through a blender to make a liquid and then boiling to destroy the enzymes. But boiling also destroys the polyurinide, rendering the gel ineffective except as a laxative. (The active ingredient in the old-time Carter's Little Liver Pills was aloin from the aloe plant.)

The polyurinide may be extracted as a powder from the aloe gel by a patented process requiring 250 pounds of aloe leaves to produce a pound of powder. It can also be obtained by a freeze drying method.

A word of caution: Do not use the whole leaf in making tea or internal preparations of any kind! The leaf contains an irritating substance especially injurious to the kidney. This substance protects the leaf from insect attacks.

(Aloes have always fascinated me. I had a lovely cluster of aloe growing in a large white cement pot outside my front door. It needed very little watering and only an occasional calcium tablet dissolved in water to stay lush and green. It had a pretty orange flower stalk when it bloomed. Now we have several beds of various types of aloes in the yard. Aloe plants are available at most nurseries and every household should have at least one. M.V.)

Winged Beans:
A Tropical Food Wonder

In 1974 the U.S. National Academy of Sciences surveyed the vast tropical plant kingdom for underexploited crops with economic value. A little known tropical legume, *Psophocarpus tetragonolobus,* commonly known as the winged bean, attracted the most attention. *Time* magazine even featured it as a "wonder crop," probably because all parts of it are edible.

The *tetragonolobus* part of the bean's botanical name tells us it has four sides, which gives the pods a fringed or winged appearance. We planted it as soon as we could get the seed from the University of Florida, which distributed seeds in the 1970s. The plant grew well during our hot, muggy summers. We tasted the edible leaves, stems, pealike blooms, young tender pods (the best part, we agreed), older beans, which tended to become tough, and the underground tubers, which remain after the plants die back. Most amazing is the nutritional content of the winged bean. It has 10 times the protein of potatoes and the equivalent protein content of soybeans. And it also enriches the soil where it grows!

The seed can be planted in rows or squares. Some reports say the plants do best in full sun. Ours flourished in partial shade. Because of their susceptibility to nematodes, the winged bean should not be planted where other types of beans were recently grown. Soil should be enriched with manures, fertilizers, compost, and worm castings.

The bean is very hard and should be scratched before planting. We use a sandpaper rub and overnight soaking. Germination is fairly low, so don't expect all of your seed to sprout. For best pod production, build a sturdy trellis for plants to climb. Pods develop in 2 stages. About 20 days after blooming, the pod forms. It grows from 2 to 14 inches. When the pods are tiny and tender, pick them, and cook as you do

Oriental snow peas pod and all.

In about 44 more days, seeds mature and pods dry and shrivel. Pick and shell and cook as you do other dried beans. Later, the underground tubers can be dug and cooked and eaten like potatoes.

From its birthplace in New Guinea, the winged bean is now grown in warm places all over the world! Recently it has appeared in gourmet produce departments. Varieties include Bogoar, Chimbu (the type we tried in Florida), Mariposa, Ribbon, Siempre, and others.

Quail Grass:
A Botanical Beauty

Contributed by Dr. Martin Pnce, ECHO Research Station, Fort Myers, Florida

Quail grass, botanically known as *Celosia argentea,* loves hot tropical summers but does well in cooler seasons too. It is an important vegetable in Africa and other developing countries. Actually it is not a grass but is in the same genus as ornamental celosia and cockscomb.

The leaves may be cooked and served like spinach. Young growing tips can be harvested, or older leaves can be used. Only a few minutes on the stove is needed to soften the leaves. Water used for cooking becomes an unappetizing black, but leaves stay green. They taste like spinach with no trace of bitterness. Discard the cooking water (you could pour it on your compost pile) as it contains oxalic acid that may interfere with calcium absorption. The oxalic acid in the leaves should pose no danger unless eaten in very large amounts.

Quail grass can be used with other greens in stir-fry dishes. Specific nutritional information is not available but as quail grass is in the same family as amaranth, food values should be similar. The leaves are probably high in vitamins A and C, iron, and calcium. (But the calcium is tied up by acid and not usable by the body.)

Quail grass grows to 8 feet if spaced about 1 foot apart. Shorter spacings are probably better in home gardens. As the days become shorter in late August, purple blooms appear. The inflorescence becomes longer and longer, remaining purple at the top. The bottom end becomes brown and contains ripe seed while blooms continue and more seeds are forming.

We have no disease problems and very little insect damage, even though we do not spray with insecticides. Because quail grass is susceptible to nematodes, a mulch is helpful. Use pine needles, leaves, and dried grass clippings. Quail grass will be killed in standing water or by freezes.

Quail grass also has potential in Northern areas because it is often hard to grow spinach in hot summer weather.

Quail grass grows as easily as weeds! Once you plant it, reseeding is done automatically.

(We especially enjoyed the colorful plumes when they were dried and used in wreaths and other flower arrangements. And it made a nice cash crop for a friend who grew and sold quail grass especially to garden club members. M.V.)

Okra:
A National Treasure

Contributed by Betty Mackey, formerly of Longwood, Florida, now living in Wayne, Pennsylvania

Although okra is one of the world's oldest and most widespread vegetables, it would still be new to most folks from the North! Okra's unusual flavor and its mucilaginous, pointed pods and creamy round seeds are quite different from most other vegetables. But cooks and gardeners are finding more and more uses for it, as well as more varieties to grow.

Okra, also called gumbo or ladies' fingers, originated in northern Africa. Young pods are rich in vitamins A and C. As the seed matures, the protein content increases to a significant amount. Okra thrives in hot areas with warm nights and is well suited to the subtropics.

The plants, when well grown, are attractive, especially the red variety, which is sometimes used in flower beds. Okra is closely related to the garden hibiscus, as can be seen in its yellowish white 2-inch blossoms with dark red throats.

Okra pods can be used in all stages of their growth. The younger ones, which contain gelatinous gum, are used whole or sliced crosswise for thickening soups, stews, and gumbo. The pods are good dipped in batter or cornmeal and fried, or steamed and served hot with butter, or cold served with salad dressing. They are often pickled, which is a good way to preserve a heavy crop. Very young red okra pods look pretty sliced raw in salads. When mature, okra pods can be too fibrous to eat. However, the nutritious seeds can be shelled, cooked, and served like beans.

In the Middle East, okra stems, leaves, and young pods are used in poultices to reduce pain. Ripe, dry seeds are roasted and ground for use as a coffee substitute.

Okra is a herbaceous annual. Home-saved seed is viable for 2 years, sometimes longer. Standard varieties are from 4 to 6 feet tall; dwarf varieties range from 2 to 4 feet. Okra may seem a little stubborn, but if you coax it by giving it the conditions it needs, you could have enough to go into the gumbo business. Once established okra can take hot, dry spells, but in order to sprout, it must be warm and wet. For fast germination, soak seeds in lukewarm water for 2 days, changing the water several times.

Plant seed ¼-inch deep, 6 inches apart, in a deep layer of prepared compost-enriched soil. It does best in full sun but will tolerate some shade in the South.

When seedlings are 4 to 5 inches tall, thin to 12 to 18 inches apart. In some areas okra attracts nematodes. You can try growing dwarf okra in large containers filled with rich, moist commercial potting soil. Clemson Spineless and Tender Pod varieties are good pot choices. Mulching your garden rows keeps weeds from sprouting.

In the subtropics, make a second planting a month or two after the first. And keep soil evenly moist to prevent bud drop. Give plants an extra boost of fertilizer about 6 weeks after they sprout.

The best time to harvest your okra is a few days after the flower petals fall. Later they become tough and fibrous. Small pods are most desired. When your okra starts to bear, experiment to see how long you can let pods grow without losing their tenderness. Pick pods frequently to keep plants in production.

Okra recipes

Cold Okra in Vinaigrette Dressing

Steam about 50 small tender okra pods just until done—about 5 minutes. Place in glass bowl and refrigerate. Mix vinaigrette dressing by putting the following into a jar:

> ¼ cup red wine vinegar
> ¼ cup salad oil
> ¼ cup olive oil
> 1 clove minced garlic
> 1 teaspoon dry mustard
> 1 teaspoon black pepper
> ½ teaspoon salt
> 1 tablespoon minced onion
> 1 tablespoon minced parsley
> ½ teaspoon tarragon
> ¼ teaspoon cayenne pepper

Cover jar with lid. Shake and store in refrigerator. Pour about half on the okra, gently mix, and let marinate in refrigerator 1 hour or more. Toss leaf lettuce with some of dressing. Line salad plates with lettuce and add marinated okra. Garnish with parsley and radishes. Serves 6 to 8.

Okra Soup

> 4 cups beef broth
> 1 pound young okra, trimmed and sliced
> 2 stalks celery
> 1 onion, minced
> 3 ripe tomatoes, peeled and chopped
> ½ cup cooked rice
> 1 bay leaf
> 1 small hot pepper, minced
> ¼ teaspoon dried or ½ teaspoon fresh thyme

Brown beef in soup pot, crumble, and pour off fat. Add all ingredients except rice and okra. Simmer 10 minutes, stirring occasionally. Add okra and rice. Stir once and simmer for 20 minutes. Serves 4.

Pigeon Pea:
Pleasing Nutritionally

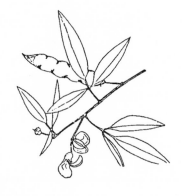

Anyone who lives in a warm climate and doesn't have pigeon peas in the garden is missing some good eating as well as easy growing. This native of Africa (*Cajanus cajan*) has been grown for centuries. A perennial, it can be a bush or small tree. Most of the many varieties will give you peas in about 7 months of frost-free weather.

Pigeon peas thrive on poor sandy soil with few if any insect problems. We've found it best to start pigeon pea seed directly in the garden after preparing the area with manures, organic fertilizers, and plenty of mulch for ground cover. We spray the leaves with fish emulsion about twice a year when we do our fruit trees.

We use the young green peas or wait until they dry. The segmented pods are hard to shell. But you can put them on cookie sheets in the sun with a screen on top and most of the pods will pop open.

Dr. Franklin W. Martin, noted tropical food researcher, reports that the leaves and young shoots are edible. The cooked leaves have a strong spicy odor, with a bit too much fiber, he says. But all parts of the pigeon pea are high in valuable protein.

Pigeon peas are used to improve the soil in many areas. When about 4 months old they are plowed under. If you live where you have occasional frost, plant your pigeon peas in a protected southern location.

Pigeon Pea recipes

Bahama Pigeon Peas and Rice

    1 cup pigeon peas (dried, soaked overnight, and
        drained) or 2 cups freshly shelled green pigeon
        peas
    1 medium onion, chopped
    2 cups cooked rice
    1 coconut or 1 cup canned coconut milk
    4 slices bacon
    1 green pepper, chopped
    2 tomatoes, chopped

Cook dried pigeon peas in 3 cups salted water (½ teaspoon salt) until tender (about 2 hours). Green peas can be cooked in just enough water to cover for about 15 minutes, or they can be microwaved on high for about 3 minutes. Fry bacon, drain and chop. Add onions, pepper, and tomatoes. Make coconut milk by pouring boiling water over grated coconut. Drain through cheesecloth and squeeze dry. Measure 1 cup, adding regular milk if necessary. Add rice, coconut milk, bacon, and vegetables to pigeon peas. Cook about 5 minutes, stirring constantly. Add a dash of hot peppersauce if desired.

Pigeon Pea Salad

    1½ cups dried pigeon peas, soaked overnight or 2
        cups freshly shelled green pigeon peas
    2 cloves garlic, chopped
    3 green onions, chopped
    4 tablespoons lemon or lime juice

Cook pigeon peas. Drain and place in bowl. Add garlic, onions, and lemon juice; mix well. Add your favorite seasoning salt to taste, or a dash of hot pepper sauce. Chill. Serve on plates of fresh garden lettuce.

Use pigeon peas with ham to make delicious soup. I'm sure you will find many ways to use your own homegrown pigeon peas! Our dear friend and former Peace Corps member, Melvin Manthey, composed this little ditty:

"Pigeon pea, pigeon pea, I sing of thee.
For I know for ere you grow.
One gets the pea.
But added too is fertility!"

Gotu-Kola:
Facts about Nature's Mystery Plant
Contributed by Dr. Dwayne Ogzewalla, Cincinnati, Ohio

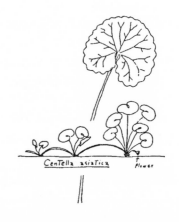

Gotu-kola, a dry herb, is sold in many health food stores but can be grown easily in a garden. Some older literature indicates that it is a caffeine-containing plant because of the "kola" name, but this is not so. It comes from *Centella asiatica* (old name *Hydrocotyle asiatica*), a little creeping plant belonging to the carrot family. Its similarity to other plants of the carrot family is difficult for nonbotanists to see.

In India, Thailand, and other tropical Old World countries, gotu-kola is used extensively as a food, being served in some of the most exclusive restaurants of Thailand. Because it contains medicinal chemicals, the safety of eating large quantities is not known. Remember that ginseng and edible burdock are toxic, if significant amounts are ingested.

Young leaves are used fresh in tossed salads; they can be wilted during the last few minutes of cooking over rice dishes; or they can be cooked in vegetable stews. In Thailand the juice of the leaves is used as a refreshing drink. The flavor of gotu-kola is aromatic with a slight bitterness, so I like it best with a spicy salad dressing or sauce.

Gotu-kola grows best in moist, shaded areas with good rich soil. It forms a thick mat, spreading outward for several feet to a yard from the original plant. It grows vigorously once established and seeds itself even in temperate areas—seedlings came up in my Cincinnati, Ohio, garden, for several years. Getting it established is sometimes a problem because seed collection is difficult and the seeds I have gathered have germinated slowly and poorly. Tiny flowers are formed close to the ground and seed develops slowly, so ripeness cannot be determined. Collecting a few hundred seeds is time-consuming.

Gotu-kola is an important medicinal plant that is listed in the Indian pharmacopeia with a variety of uses. It is claimed to act on the genitourinary tract to reduce irritation, increase urine flow, and promote menstruation. It is used to reduce nervousness, and in large doses, it acts as a narcotic, produces headache, and causes giddiness and coma. It is also used as a nonspecific "tonic."

Its only well-verified use is as an aid in the healing of various skin diseases including psoriasis and leprosy. The active chemical appears to be Asiatic acid. I have been testing it for "healing" of skin irritations on myself and animals with promising results. While I have not developed an ideal formula for a lotion, I am sure that the juice of the leaves could be squeezed onto irritated skin with good results. Reports of "intelligence-expanding" and "total health-expanding" properties are mainly the advertising exaggerations of Western sellers with no clinical evidence.

Raising Exotic Quail | 11

Contributed by Dr. Elizabeth Nutting, Melbourne Village, Florida

I've never had a canary, but I have always thought it would be fun to have one—until a few years ago when I began taking care of a friend's Japanese quails. Now I say, forget about having songbirds in your house—get these miniature quails, which will give you an egg almost every day and entertainment all the time. They require minimal effort and attention—about 10 minutes, morning and evening. I feed and water the quails, change the paper in the bot-tom of their cages, and collect the eggs. I add their droppings to our earthworm boxes.

Formally called *Coturnix coturnix japonica,* these quails have been raised by the Japanese for centuries. The eggs are about one-third the size of a hen's. They are speckled brown on a tan surface. Some even look greenish. Our friends have been surprised to be served hard-boiled tiny quail eggs. Everyone usually shells his own because it can be quite a chore to peel enough for a party. Quail are also raised for their fine flesh, but I could never bring myself to eat my little friends.

The quails are fed young turkeys' feed. I add chopped sprouted mung beans, comfrey, or young grape leaves picked in the woods. I'm sure it keeps them healthy, as they really gobble it up.

Our subtropical weather is perfect for the quails. I don't have to worry about extra protection on a few cold nights, even though they are kept on a screened porch. Health problems are few, but their life span is only about 2 or 3 years.

Our 6 hens and one little cock all have different personalities. Every morning the cock rewards us with a tiny crow, and every evening the hens each make a "whowie" sound before laying their eggs.

(We were introduced to the Coturnix quail when Dr. Jerome Keuper began a research project at Florida Institute of Technology in Melbourne in 1972. He found that raising the quails was easy, but marketing the tiny eggs wasn't economically feasible. The beauty of raising Japanese quails is that they incubate in 15 days and start laying eggs at 5 weeks. In Japan, surplus males are used for meat at only 5 weeks of age. The June 1975 issue of Organic Gardening magazine featured an article about these little quail. You may be able to find this back issue in a local library. Our original California source for these quail no longer exists. Look for them in farm magazines. M.V.)

Serving Up the Bounty | 12

We have included standard or classic recipes for many of the fruits in previous chapters. They appeared in our newsletters from 1975 to 1988. Don't hesitate to reduce the sugar and fat amounts in these recipes. But because many folks are concerned about too much sugar in their diets, we offer here some ways to preserve your tropical bounty with fewer calories. (Note: When 1 box of commercial pectin is called for in the jelly recipes in the text, it is the 1¾-ounce-size box available in the United states.)

Microwave Fruit Jam

> 1½ cups mashed mango, persimmon, papaya,
> Surinam cherry, or fig
> Juice of ½ lemon
> ¼ cup sugar

Use at least a 6-cup micro-safe glass cooking dish. Mix ingredients and cook on high for 6 to 10 minutes or longer until mixture thickens. You will have to stop often, take out the dish, and stir several times to keep jam from boiling over. It will "sheet off the spoon" and thicken when done. Time will vary with your particular oven.

Pour into a large jelly jar you have rinsed out with boiling water. Make sure you pour boiling water over the cover too. Cover and cool. Then store in your refrigerator no more than two weeks. I have also made this recipe using Key lime juice, calamondin puree, and honey with fine results.

Other Low-Sugar Preserves

I recently tried Pomona's Universal Pectin, which allows you to make preserves with less sugar or honey and shorter cooking times. You can find this product at some specialty shops and health-food stores. See Gardener's Supply in Appendix 5. Besides the pectin, the box contains a separate portion of monocalcium phosphate for use with fruits especially low in calcium and hard to jell. It may be the answer to making a cactus pear jelly with less sugar. Here are two recipes from the package:

Cooked Berry Jam with Honey

> 4 cups mashed or pureed berries (strawberries, raspberries, blackberries, etc.) mixed with 4 tablespoons lemon juice
> ½ cup honey
> 2 teaspoons pectin powder mixed into paste with 1 teaspoon honey
> 2 teaspoons calcium-water solution

Heat fruit and lemon juice to rolling boil. Add honey and pectin. Boil 1-2 minutes. Remove from heat. Stir in calcium-water solution. Blend well. Pour into sterilized jars and seal.

Wild Grape Jelly

> 3 cups grape juice made from 6 cups of grapes
> 1 cup honey or sugar
> 1 teaspoon pectin powder as paste mixed with honey or sugar

Bring grape juice to rolling boil. Add pectin and sweetener. Boil 1 minute, stirring constantly. Remove from heat, pour into sterilized jars, and seal.

Using Tropical Fruit Preserves

What can you do with leftover tropical fruit preserves besides eating it on your breakfast toast or putting it on peanut butter sandwiches?

1. Try topping a roast or broiled chicken with mango chutney. Use mango and carambola preserves as glaze on baked ham. Surinam cherry jams add bright red color for garnishing fish and fowl.

2. Use some of your tart jellies (grape, Java plum, and others) melted and mixed with a little vinegar as a Chinese-type sweet-and-sour duck sauce for serving over egg rolls and pork chops.

3. Now that many of us have forsaken bacon-and-egg breakfasts for oatmeal, we can liven up that dish of porridge with a spoonful of whatever preserve we have on hand. How does red roselle jelly sound? It looks pink and pretty on top of that utilitarian cereal.

4. I received one of the new automatic bread making machines for a birthday gift. And I've had fun adding orange marmalade, carambola jam, or mango chutney—only a tablespoonful worked fine for my machine. Our breads are uniquely flavored!

5. You can add some of your tropical treats to icing for cakes, cake fillings, and topping for ice cream. Where else could you get a carambola sundae?

6. Another special machine that helps us eat well using tropical fruit is our "Italian Ice Cream Man" made by El Galato. Actually, it is a little freezer. You put your ice cream mix in the container, press the but-

ton, and 20 minutes later, you have a tasty soft fruit ice or ice cream. No messing around with ice and salt! Here are some of our favorite frozen combinations:

2 cups fresh tangelo or grapefruit juice and 2 cups apple cider (with or without a shot of vodka or rum).

2 cups papaya, persimmon, or mango puree with 2 cups apple juice or apple cider.

Papaya or banana (mashed) with an instant vanilla pudding mix; add milk to equal 4 cups.

If using a refrigerator freezer, you will have to take the tray of fruit ice out of freezer when it is half-frozen. Stir it well or beat it with a mixer. Then return it to the freezer until frozen through. This prevents large ice crystals from forming.

What to do with your leftover tropical fruit? Take care of your health and try this recipe from Willa Davidsohn of Melbourne, Florida:

Special Oat Bran Muffins

2½ cups oat bran
½ cup flour
¼ to ½ cup sugar or honey
2 tablespoons cooking oil
2 egg whites
1 cup pureed fruit
½ cup chopped walnuts

Mix dry ingredients, including nuts. Mix wet ingredients together and pour into dry. Gently mix only a few times. Pour into greased muffin tins and bake at 350 for about 20 minutes. Willa uses her homegrown calabaza, calamondins, persimmons, bananas, and other fruit for these tasty, healthful treats.

Making Juice from Tropical Fruit

1. **Micro Method:** Here's a good way to make juice when you have only a few guavas, loquats, or similar fruit ripening at a time. Pick and wash fruit. Slice into microwave dish and cover with water. Put cover on dish. (I never use plastic or wax paper). Cook on high 10 minutes. Let cool. Strain into pitcher. Press fruit against side of strainer if you want more pulp. Add sugar or honey if necessary. We hardly ever add sweetening and enjoy the juice over ice or put in freezing containers leaving 2 inches of air space on top. Date and store in deep freeze. When making party punches you can add almost any flavor you have on hand to citrus juices to get fresh fruit flavor combined with exotic tastes. A favorite of mine is fresh-squeezed grapefruit juice with guava.

2. **Steam Juicer:** I was introduced to this handy utensil when we spent summers in British Columbia and a friend let me borrow her steamer. It looks like a giant double boiler with 3 sections. Fruit goes in the perforated top basket. The center pot catches juice and runs it out a rubber tube into sterilized bottles. The bottom contains boiling water. "Make sure and don't let the water run dry in the bottom pot. The pot could burn!" my friend cautioned. We used it with apples, apricots, cherries, and grapes. When I got my own steam juicer I used it in Florida. It worked well with cactus pears, loquats, surinam cherries, as well as grapes.

3. **Power Juicers:** These are probably the best juicers as the fruit doesn't have to be cooked and more vitamins and minerals are saved. The only problem is the cost which runs into several hundred dollars. But my 103-year-old neighbor drank fresh juices—mostly vegetable—made by a power juicer for years. Need we say more?

4. **Boiled fruit juices:** Just cut fruit into a large kettle. Cover fruit with water and bring to a boil. Simmer fruit until soft. Pour it through a cloth bag. Old-fashioned flour and sugar bags work well, or make

one from sturdy cotton. If the fruit has many seeds you can run the pulp through a colander to remove them. You can use the resulting puree in many recipes besides enjoying the juice from the fruit!

Wine from Tropical Fruit

There is an ongoing debate about using chemicals in making wine. Some wine makers feel it is necessary to use Campden tablets (sodium metabisulfite) to kill unwanted native yeasts. I used it for a while, especially to sterilize my bottles. But now I wash the gallon jugs with dishwashing soap and water and rinse them with household bleach and more water, and it seems to do just as well and costs less.

Some wine makers also use a yeast nutrient, acid blend, peptic enzymes, grape tannin, and other ingredients. These may be needed when you mash up the fruit and let it set at room temperature in our hot climate. But we skip that step and begin with juice. I tried making mango wine in a 5-gallon glass jug and found it too heavy to lift. Now we make all our wines in 1-gallon glass jugs and I can swing them around easily! Here's our basic recipe, which we have tried with most of the fruit mentioned in this book. Carefully clean glass jug. My nurse friends would say, "Use sterile technique!"

Basic Wine Recipe

Dissolve 4 to 5 cups sugar in 8 cups hot water and pour into jug. Add 4 to 5 cups fruit juice.

Soften a package of regular baker's yeast (or wine yeast) in a little lukewarm water. When fruit-juice mixture is lukewarm pour yeast mixture into jug.

Put original cap on jug and shake carefully to mix. Add enough water to fill jug 2 inches from the top. (I've been using filtered city water with no problem.)

Then put on air lock and allow to ferment. It's best to keep the jug in a cool, dark place. We have a wine cellar under a stairway. We have added cinnamon and cloves to some of the juices for a spicy flavor. Use a cloth bag to hold spices when you boil the fruit to extract the juice. Or simmer an already prepared spice bag which looks like a tea bag. This works very well with tangerine and elderberry.

When bubbles stop, take off air lock and put back original jug cap. You can sterilize wine bottles and decant wine into them. We have been too busy to bother with small bottles, so we just leave the wine in the gallon jugs until ready to use. Then we carefully decant wine into vodka bottles (which are already sterilized from the original contents).

We have taken our wines made this way to several "Wild Food Weekends" and writers' conferences and folks often come back for several samples.

Of course you can get a pH meter and alcohol meter and add sugar according to the readings. But most folks are too busy for that. We've had good results with all fruit except sea grape, which turned to vinegar. But no loss as we used it for making mango chutney!

Citrus Specials

My sister-in-law, Anne Van Atta from Birmingham, Alabama, makes this Deep South longtime favorite.

Ambrosia

Peel enough oranges for the crowd. Section and, if preferred, remove white membrane. Sprinkle with powdered sugar and fresh grated coconut. You may add pineapple if desired. Serve in pretty glass bowl.

Citrus rinds may be dried in home dehydrators, home ovens, or microwaves. Follow directions for individual dehydrators, but pick a sunny day. Spread a thin layer of shredded orange or grapefruit peel on cookie sheet. Bake at your oven's lowest setting with the door cracked open for about an hour or until on pinching fruit between fingers, you do not feel any moisture.

In microwave put shredded peel on paper towel on top of flat glass dish. Cover with another sheet of paper towel. Experiment with your oven to see what temperature and time work best. Be careful not to burn peel. You may have to rotate plate.

Dried peel can be used for your own homemade tea mixtures. Look at the boxes of herbal teas for ideas. We combine orange and lemon peel with roselle, mint, and broken cinnamon sticks. Dried peel can also be added to cookie dough. Citron is the secret ingredient in hermit cookies. Of course, homemade fruit cakes use citron and other peels too.

Fresh citrus rinds are fun to use as garnishes. You can make orange, lemon, lime, or grapefruit roses by rolling up strips of peel or you can make chrysanthemum flowers by carefully cutting the peel with a sawtooth cut around the middle, leaving the bottom connected. Or just use thin slices of any fruit. You can flute the edges using a little knife called a zester.

The tiny, sour calamondin is ideal in cakes and breads. When your tree is loaded with fruit, I'm sure you will want to make this cake. I won first prize for it in a Florida Heritage Recipe Contest.

Mother Nature's Florida Calamondin Supreme Cake Mixture

½ cup calamondin puree
1 package plain white cake mix
1 3-ounce package lemon Jell-O
⅓ cup milk
4 large eggs
¾ cup cooking oil
1 tablespoon fresh Florida lemon juice

Make puree by cutting calamondins in half, removing seeds, then cutting in quarters and running in blender. Mix cake mix, dry Jell-O, milk, and calamondin puree together. Beat eggs, oil, and lemon juice together. Add to cake mixture. Pour into oiled and floured 10 x 4½-inch angel food or Bundt cake pan. Bake in a preheated 350 degree oven until a sharp knife comes out clean. Carefully remove from pan onto cake platter. While still warm spread with glaze made by mixing together:

> ½ cup calamondin puree
> 4 tablespoons butter or margarine
> 1 tablespoon fresh lemon juice
> 2 cups powdered sugar

Spread glaze over cake. Place a pecan half over each serving. Garnish with thinly sliced calamondins or calamondins cut into flowers and fresh green mint leaves.

This cake can also be baked in a 13x9x2 inch pan and is less work to prepare. Rare fruit clubs in Vero Beach, Sarasota, and Fort Myers have served it when we gave programs at their meetings. If you are watching calories, make calamondin cake by using only 2 eggs and 1⅓ cups water instead of oil. You can even skip the the Jell-O as the calamondin flavor is very pronounced. (I have also substituted guava or passion fruit puree for the calamondin with fine results.)

One of the best fruit breads you can make with calamondins is this recipe from Chris Bruton, Merritt Island, Florida, a longtime writer friend. It appeared in issue No. 72 of our "Living off the Land" newsletter.

Chris Bruton's Calamondin Bread

Wash and remove seeds of calamondins. Use both skin and pulp and blend to a smooth paste.

Mix and set aside:
> 1 cup calamondin paste
> ⅓ cup sugar

In a separate bowl mix:
> 1 stick margarine
> ¾ cup brown sugar
> 2 eggs

Sift together in separate container:

> 2 cups flour
> 1 teaspoon salt
> 1 ½ teaspoon baking soda
> ¼ teaspoon cloves
> ½ teaspoon cinnamon

Blend together alternately with paste into margarine and sugar mixture. Then add ¾ cup chopped pecans or walnuts and ¾ cup raisins. Pour into greased and floured loaf pan and bake at 350 for 50 to 60 minutes.

Key Lime recipes
Contributed by Vanette Skillin, Bonita Springs, Florida

Usually some type of lime, including our hardy Lakeland lime which is of half Key lime parentage, is available from our yards. Store excess juice by freezing it in ice cube trays. Pop cubes out of trays and keep in plastic freezer bags.

Old Sour
This is our favorite condiment, which we use on avocados, fish, salads, and even cottage cheese.

> 1 cup Key lime juice (can also make this with lemon
>   and sour orange juice but flavor is slightly
>   different)
> 1 tablespoon salt
> 2 whole bird peppers or a few drops hot pepper sauce

Combine all ingredients and pour into sterilized bottle. Shake to dissolve salt. Lightly cork and let stand at room temperature on your kitchen counter until fermented, usually 2 or 3 days during a Florida summer. Then replace original cover and store in refrigerator. Make sure all fermentation is finished or it may blow up!

Key Lime Daiquiri

Don't bother to buy daiquiri mix when you grow your own limes.

> 2 limes, juiced
> 2 ounces white or dark rum
> 5 to 6 tablespoons sugar (Use raw if you can find it.)

Put lime juice, sugar, and rum in blender. Use a blender which will crush ice and add ice cubes and rum until mixture is of sherbet consistency. Spoon into chilled glasses and garnish with fresh mint.

We were served this fine drink at the Tijuana Art Museu when we visited Mexico with Rare Fruit Club members.

No-Egg Key Lime Pie

> ⅓ cup freshly squeezed Key lime
> 1 can Eagle brand sweetened condensed milk
> ¼ teaspoon almond extract juice
> ½ cup Cool Whip or other topping (optional)

Combine all ingredients, folding in the whipped topping. Pour into baked 9-inch pie shell and top with more Cool Whip. Chill and serve garnished with thin lime slices and fresh mint leaves.

Easy Elegant Desserts with Tropical Fruit

Use the sponge cake bases and the fresh crepes now found in produce departments to create works of art with your homegrown tropical fruit. I use instant pudding mix to cover the top of a sponge cake ring. Then I place sliced carambolas, guava rings, persimmon slices, and mango, kiwi, and papaya slices in circles around the top. Next I add a thin covering of glaze made from homemade preserves such as guava, Surinam cherry, or loquat jelly. Jelly can be melted in your microwave—just cool it a bit before covering fruit. Top the fruit with vanilla yogurt or a dairy whip.

For a low-calorie treat, I serve sliced peaches, mangos, papayas, bananas, tropical cherries, blackberries, mulberries, figs and other fruit spread on a crepe, rolled up and garnished with a slice of fresh fruit on top of a dab of vanilla yogurt.

When you grow your own exotic fruit, I am sure you will find many ways to serve them and enjoy their fresh, healthful benefits!

Rare Fruit Grower and Related Organizations

For trees, plants, and information, join a rare fruit grower club or a related organization. Because most clubs are run by volunteers from their homes, and officers change yearly, no phone numbers are given. Clubs usually keep the same address for many years. Note that clubs marked RFCI are associated with the Rare Fruit Council International.

Rare Fruit Council International
Box 56194
Miami, FL 33256

The following 6 clubs are associated with this founding association:

Brevard (RFCI)
P.O. Box 3773
Indialantic, FL 32903

Indian River (RFCI)
P.O. Box 2338
Vero Beach, FL 32961-1117

Manatee (RFCI)
P.O. Box 1656
Bradenton, FL 34206

Palm Beach (RFCI)
P.O. Box 16464
Palm Beach, FL 33416

Sarasota Fruit & Nut Society (RFCI)
4520 Camino Real
Sarasota, FL 33581

Tampa Bay (RFCI)
P.O. Box 20636
Tampa, FL 33685

Other Clubs:
Caloosa Rare Fruit Exchange
Box 3406 Palm Beach Blvd.
Fort Myers, FL 33905

Collier Fruit Growers Council
P.O. Box 9401
Naples, FL 33941

Rare Fruit and Vegetable
Council of Broward County
3245 S.W. 70th Ave.
Ft. Lauderdale, FL 33312

Southwest Florida Rare Fruit
Growers Exchange, Inc.
P.O. Box 8923
Naples, FL 33941

Tropical Fruit and Vegetable
Society of the Redland
c/o Fruit & Spice Park
24801 S.W. 187 Ave.
Homestead, FL 33031

California Rare Fruit Growers
Fullerton Arboretum
California State University
Fullerton, CA 92634
(Will give you locations of rare fruit
    groups in California)

Pits
Debbie Peterson
251 11th St.
New York NY 10014
(Especially helpful if you want to start
    fruit-bearing plants from seed and
    keep them growing in pots.)

Israel Rare Fruit
Ariel Shai, Horticultural
Research and Development
Tropical, Subtropical Rare Fruits
Jacobson 5 St.
Rehovot
Israel 76206

North American Fruit Explorers
P.O. Box 94
Chapin, IL 62628

Rare Fruit Council of Australia
P.O. Box 707
Cairns, Queensland 4870
Australia

New Zealand Tree Crops Association
P.O. Box 1542
Hamilton
New Zealand

Seed Sources

There is always a danger of importing disease with seed and plants. If you order seeds from overseas, you must obtain an import permit. For an application write to:

Permit Unit
USDA-APHIS-PPQ
665 Federal Building
Hyattsville, MD 20782

Southern Seeds
P.O. Box 2091
Melbourne, FL 32902
Send S2.00 (refundable) for catalog. (This company carries the Marian Van Atta Calabaza seed.)

John Brudy Exotics
3411 WestField Dr.
Brandon, FL 33511
Send $2.00 (refundable) for catalog.

Echo, Inc.
17430 Durrance Road
North Fort Myers, FL 33917
Also gives tours. Needs volunteer workers. Grows food-producing seeds for developing countries but has some available for sale. Send $2.00 and a large self-addressed, stamped envelope for information.

Kilgore Seed Company
1400 W. First Street
Sanford, FL 32711
Has many hard-to-find items for southern growers, including chuf-fa and special peanut varieties. Write for catalog.

Park Seed Company
Cokesbury Road
Greenwood, SC 29647-0001
These folks sent seeds into outer space and distributed tomato seed to schools for test planting. They have a great selection of seed for herbs, vegetables and flowers for warm places. Write for catalog.

Hastings
P.O. Box 4274
Atlanta, GA 30302-4274
Has southern fruit and nut trees, berries, and more. Write for cata-log.

Australian Seed Source
Fruit Spirit Botanical Garden
Dorroughby 2480
New South Wales
Australia
Write for their extensive seed list.

Israel Seed Source
Natan Vardi & Son
50297 Hashivah
Israel
Write for their interesting seed list.

3 | Fruit Tree Sources

Florida

Rockledge Gardens
2153 South U.S. Highway 1
Rockledge, FL 32955
(321) 636-7662
Has a fine selection of high-quality
fruit trees for the subtropics.

Nelson's Landscape & Nursery
P.O. Box 12395/1085 John Rodes
Blvd.
West Melbourne, FL 32912-0395
Nelson's has been very helpful with
our garden problems through the
years.

Sun Harbor Nursery
930 E. Eau Gallie Blvd.
Indian Harbour Beach, FL 32937
(321) 773-1375
Helped us landscape two houses
with edible ornamentals. Owner,
David Grover, has written for our
newsletter.

Downtown Produce
875 Creel Street
Melbourne, FL 32935
(321) 724-1242
Carries tropical fruit trees and herb
plants as well as many fresh tropical fruits. In past years we have
enjoyed bunchiosa, black sapote,
canistel, and more from this store.

Fellsmere Nurseries
11280 138th Avenue
Fellsmere, FL 32948
(561) 571-1069
Located in the old sugar-mill town
of Fellsmere, this nursery has
developed into a source for hard-
to-find tropical items.

Our Kids Tropicals & Nursery
17229 Phil C. Peters Rd.
Winter Garden, FL 32787
Has an outstanding collection of
tropical fruit plants and trees,
including many banana varieties
and even miracle fruit.

Chestnut Hill Nursery
15105 NW 94th Avenue
Alachua, FL 32615
(904) 462-2820
A good mail order source for chestnuts and oriental persimmons.

Szabo Nursery Garden Center
15450 New U.S. 41
Naples, FL 33963
(941) 597-2322
Has an outstanding collection of
tropical food plants, beautifully
displayed.

Ward Reasoner & Sons Landscaping
5855 45th St. East
Bradenton, FL 34203

(941) 756-1881

One of the very first nurseries in Florida, Reasoner & Son's has many hard-to-find items as well as grafted hibiscus.

Hopkins Citrus & Rare Fruit Nursery
5200 SW 160th Avenue
Fort Lauderdale, FL 33331

Has an outstanding collection of tropical fruit trees from akee to sapodilla and much more.

Zill High Performance Plants, Inc.
6801 107th Pl. S.
Boynton Beach, FL 33437
(561) 732-3555

Carries many avocado and mango varieties, including the dwarf, Julie.

California

Atkins Nursery Citrus
3129 Reche Road
Fallbrook, CA 92028
(760) 728-1610

Specializes in avocado and citrus, plus many subtropical and deciduous fruit trees.

Pacific Tree Farms
4301 Lynwood Dr.
Chula Vista, CA 91910
(619) 422-2400

They have black cherry (Capulin), several warm-weather raspberries, and much more.

Papaya Tree Nursery
12422 El Oro Way
Granada Hills, CA 91344
(818) 363-3680

Owner David Silber wrote an article for our newsletter about the exciting Babaco, a mountain papaya which likes some cold weather.

Mail Order Produce and Cooking Supplies

Star Organic Produce
P.O. Box 551745
Fort Lauderdale, FL 33355
(305) 262-1242
Has organically grown citrus, avocados, mangos, papayas, bananas, and
more.

Frieda's by Mail
4465 Corporate Center Dr.
Los Alamitos, CA 90720
(800) 421-9477
Has an exotic fruit basket available
mail order, which includes many
of the items mentioned in this
book. You could eat the fruit and
save the seeds to plant.

Stark Brothers Nurseries
P.O. Box 10
Lousiana, MO 63353-0010
(800) 325-4180
This long-time company has tropical
kiwis, Asian pears, and apples for
zone 9. Write for catalog.

Gardener's Supply Co.
128 Intervale Rd.
Burlington, VT 05401
(802) 660-3505
Has Pomona's Universal Jam-Jelly
Pectin for sugar-free preserves.
Also carries steam juicers.

Classes, Lectures, and Places to Visit and Learn

The rare fruit societies listed in Appendix 1 all have informative programs at their regular meetings. In addition, several have demonstration fruit orchards. Here are some other places of interest in Florida and California.

East Central Florida

Gleason Park
2055 S. Patrick Dr.
Indian Harbor Beach, Florida
(321) 773-0552
The Brevard Rare Fruit Council landscaped this park with tropical fruit-producing plants. It is open to the public daily.

Florida Institute of Technology Botanical Garden
150 West University Boulevard
Melbourne, Florida
(321) 674-8000
Open daily from dawn to dusk. The Friends of the Botanical Garden sponsor a yearly plant ramble.

Beachland Elementary School
3551 Mockingbird Dr.
Vero Beach, Florida
(561) 564-3300
The Indian River Rare Fruit Council has landscaped the campus of this school with many tropical edibles. They are used to teach the children about tropical agriculture. The fruit club gives guided tours.

Central Florida

Herry P. Leu Gardens
1920 North Forest Avenue
Orlando, FL 32803-1903
(407) 246-2620
Offers classes and a yearly garden show.

South Florida

Flamingo Gardens
3750 Flamingo Road
Fort Lauderdale, FL 33069
(954) 473-2955
Has tropical plants, garden events and classes.

Mounts Botanical Garden
531 Military Trail
West Palm Beach, FL 33415
(561) 233-1749
Many tropical fruit trees produce their unusual crops at this outstanding educational facility and center for urban horticulture.

Miami-Dade Community College
Environmental Center
11011 SW 104th Street
Miami, FL 33276-3393
(305) 237-0905
Offers classes in landscaping, cooking, building for a warm climate, and much more.

Fairchild Tropical Garden
10901 Old Cutler Road
Miami, FL 33156
(305) 667-1651
This garden houses an outstanding palm collection as well as many tropical fruit trees. Has a wide selection of classes and a yearly ramble, including a plant sale.

Fruit & Spice Park
24801 SW 187th Avenue
Homestead, FL 33031
(305) 247-5727
This former farm has many producing tropical fruit trees, wild edible classes, and field trips. It is our favorite place to visit.

West Coast Florida

Conservancy of SW Florida at Naples
1450 Merrihue Drive
Naples, FL 33942
(941) 262-0304
Offers classes and special events. I gave a day-long "Living Off the Land" seminar here some years back.

Marie Selby Botanical Gardens
811 S. Palm Avenue
Sarasota, FL 34236
(941) 366-5730

Has displays, events, special classes, and a wonderful tropical vegetable garden.

Eden Vineyards
19709 Little Ln.
Alva, FL 33920
(921) 728-9463
Features herb classes, wine, a bakery, cheese bar, and more. A visit to this beautiful vineyard will give you much pleasure.

Sunburst Tropical Fruit Park
7113 Howard Road
Bokeelia, FL 33922
(813) 283-1200
Call to schedule a tour of the park and taste tropical fruit preserves and fresh mangos.

California

The Quail Botanical Gardens
230 Quail Gardens Dr.
Encinitas, CA 92024
(760) 436-9466
Includes the oldest collection of subtropical fruits on public display in the United States. See Surinam cherries, macadamia trees, strawberry guavas, and more.

Demonstration Fruit Orchard
12601 Mulholland Dr.
Beverly Hills, CA 90210
(310) 285-2537
Located just off the intersection of Coldwater Canyon and Mulholland Drive, the orchard is found at the end of the upper nature walk on Tree People grounds. Call the City of Beverly Hills for more information.

Bibliography

Wili, E. Golby and L. Maxwell. *Florida Fruit*. Tampa, Florida, Lewis S. Maxwell, Publisher, 1967, 120 pages.

Harrison, S. G., G. B. Masefield and M. Wallis. *Oxford Book of Food Plants,* London, England, Oxford University Press, 1969, 206 pages.

Morgenstern, Jayne H., compiler. *Cookbook,* Pompano Beach, Florida, The Rare Fruit and Vegetable Council of Broward County, 354 pages. (For availability and price, contact this organization at 3245 SW 70th Ave., Ft. Lauderdale, FL 33313.)

Rare Fruit Council International, Inc. *Tropical Fruit Recipes,* Miami, Florida, Rare Fruit Council International, 1981, 180 pages.(Contact the council at Box 56194, Miami, FL 33256, for current availability and price.)

Staff of the L. H. Bailey Hortorium, Cornell University. *Hortus Third.* New York, Macmillan Publishing Co., Inc., 1976, 1290 pages.

Sturtevant, E. L. *Sturtevant's Edible Plants of the World, 1919,* New York, Dover Publications, Inc., 1972, 686 pages.

Index

I f you enjoyed reading this book, here are some other Pineapple Press titles you might enjoy as well. To request our complete catalog or to place an order, write to Pineapple Press, P.O. Box 3889, Sarasota, Florida 34230, or call 1-800-PINEAPL (746-3275). Or visit our website at www.pineapplepress.com.

Essential Catfish Cookbook by Shannon Harper & Janet Cope. Mouth-watering recipes that call for succulent catfish and a variety of easy-to find ingredients. Learn about the private life of the captivating catfish and enjoy this Southern delicacy. ISNB 1-56164-201-0 (pb)

The Everglades: River of Grass, 50th Anniversary Edition by Marjory Stoneman Douglas. This is the treasured classic of nature writing that captured attention all over the world and launched the fight to save the Everglades when it was first published. The 50th Anniversary Edition includes an update on the events in the Glades in the last ten years. ISBN 1-56164-135-9 (hb)

Growing Family Fruit and Nut Trees by Marian Van Atta. What better way to celebrate your family than by growing a tree whose delicious fruit will be a yearly reminder of important events? Learn to choose the right trees and to keep them healthy and bountiful. ISBN 1-56164-001-8 (pb)

Guide to the Gardens of Florida by Lilly Pinkas. Organized by region, this guide provides detailed information on the featured species and facilities offered by Florida's public gardens. Includes 16 pages of color photos and 40 line drawings. ISBN 1-56164-169-3 (pb)

Mastering the Art of Florida Seafood by Lonnie T. Lynch. Includes tips on purchasing, preparing, and serving fish and shellfish—with alligators thrown in for good measure. Includes tips for artistic food placement, food painting techniques, and more. ISBN 1-56164-176-6 (pb)

The Mongo Mango Cookbook by Cynthia Thuma. Features mangoes in salads, meat and seafood dishes, desserts, drinks, and even salsas and chutneys. An appealing blend of Asian, Mexican, Indian, and American recipes awaits! Includes mango history and lore. ISBN 1-56164-2398 (pb)

The Mostly Mullet Cookbook by George "Grif" Griffin. Includes dozens of mullet main dishes, such as Dixie Fried Mullet, Mullet Italiano, Sweet & Sour Mullet, and the Sea Dog Sandwich, as well as mullet-friendly sides and sauces and other great Southern seafood, including Judy's Mullet Butter and Ybor City Street Vendor's Crab Cakes. ISBN 1-56164-147-2 (pb)

The Sunshine State Cookbook by George S. Fichter. Delicious ways to enjoy the familiar and exotic fruits and vegetables that abound in Florida year-round. Includes seafood cooking tips and delectable ideas such as Rummed Pineapple Flambé and Caribbean Curried Lobster. ISBN 1-56164-214-2 (pb)

Visiting Small-Town Florida Volumes 1 and 2 by Bruce Hunt. Filled with both color and black-and-white photos, these two guidebooks will have you exploring the Florida that used to be. Visit charming inns, antique shops, and historic homes, and sample authentic home cooking along the way! Volume 1: ISBN 1-56164-128-6 (pb); Volume 2: ISBN 1-56164-180-4 (pb)